Contents

Chapter 1
Preliminaries

Time-delay systems are modeled by ordinary differential equations which involve delayed variables (Gu et al. 2003; Krstic 2009). They are frequently encountered in many applications, for instance, in biology or biomedical systems (Timmer et al. 2004), in telerobotics, teleoperations (Islam et al. 2013; Kim et al. 2013), and in networked control systems.

The literature on time-delay systems is extensive, and concerns mainly stabilization problems. The analysis of the structural properties of this class of systems is less developed, both with respect to the delay-free case and the linear case. Even fundamental properties such as accessibility or observability and related design problems are far from being completely understood.

The aim of this book is to introduce the reader to a new methodology, recently introduced in the literature, which has allowed to obtain interesting results in the analysis of the structural properties of time-delay systems affected by constant commensurate delays. In order to give the reader a flavor of the important issues that arise in the delay context, we give hereafter an overview of the different problems that will be analyzed later on.

Such overview will show that the results in this book feature fundamentals of a novel approach to tackle nonlinear time-delay systems. They include useful algebraic results which are independent of any system dynamics.

1.1 The Class of Systems

Let us for the moment focus our attention on the class of single-input nonlinear time-delay systems which can be described through the ordinary differential equation

$$\dot{x}(t) = F(x(t), \ldots, x(t - sD)) + \sum_{i=0}^{l} G_i(x(t), \ldots, x(t - sD))u(t - iD)), \quad (1.1)$$

where $x(t) \in \mathbb{R}^n$ and $u(t) \in \mathbb{R}$ represent, respectively, the current values of the state and of the control; D is a constant delay; $s, l \geq 0$ are finite integers; and finally the functions $G_i(x(t), \ldots, x(t - sD))$, $i \in [0, l]$, and $F(x(t), \ldots, x(t - sD))$ are analytic in their arguments. It is easy to see that such a class of systems covers the case of constant multiple commensurate delays as well (Gu et al. 2003).

Without loss of generality, in order to simplify the notation, we will assume $D = 1$. Then, system (1.1) reads as follows, for $t \geq 0$:

$$\dot{x}(t) = F(x(t), \ldots, x(t - s)) + \sum_{i=0}^{l} G_i(x(t), \ldots, x(t - s))u(t - i)). \quad (1.2)$$

The differential equation (1.2) is influenced by the delayed state variables $x(t - i)$, $i \in [1, s]$, which, for $t > s$, can be recovered as solution of the differential equations obtained by shifting (1.2), and thus given for $\ell \in [1, s]$ by

$$\dot{x}(t - 1) = F(x(t - 1), \ldots, x(t - s - 1))$$
$$+ \sum_{i=0}^{l} G_i(x(t - 1), \ldots, x(t - s - 1))u(t - i - 1))$$
$$\vdots \quad\quad\quad\quad\quad\quad\quad\quad (1.3)$$
$$\dot{x}(t - s) = F(x(t - s), \ldots, x(t - 2s))$$
$$+ \sum_{i=0}^{l} G_i(x(t - s), \ldots, x(t - 2s))u(t - i - s)).$$

The extended system (1.2), (1.3) now displays s additional delays on the state and input variables. Virtually, one can continue this process adding an infinite number of differential equations which allow to go back (or forward) in time, and thus leading naturally to an infinite dimensional system.

As well known, from a practical point of view, time-delay systems are at a certain point initialized with some arbitrary functions, which may not necessarily be recovered as a solution of the differential equations describing the system. The consequence is that any trajectory of the extended system (1.2), (1.3) is also a trajectory of (1.2), but conversely any trajectory of (1.2) is, in general, not a trajectory of (1.2), (1.3).

Example 1.1 Consider the scalar delay differential equation

$$\dot{x}(t) = -x(t - 1) \quad\quad\quad\quad (1.4)$$

defined for $t \geq 0$. To compute its trajectory, it requires a suitable initial condition, say the function $\varphi_0(\tau)$ defined for $\tau \in [-1, 0)$.

The addition of the extension

$$\dot{x}(t - 1) = -x(t - 2) \tag{1.5}$$

implies that with the initial condition $\varphi_0(\tau)$ on the interval $\tau \in [-1, 0)$, the system (1.4), (1.5) is well defined for $t \geq 1$. Alternatively, if (1.4), (1.5) is considered for $t \geq 0$, then an additional initial condition $\varphi_1(\tau)$ should be given on the interval $\tau \in [-2, -1)$. However, it should be noted that the trajectory of (1.4) corresponding to the initial condition $\varphi_0(\tau)$ cannot be reproduced, in general, as a trajectory of the extended system (1.4), (1.5), as there does not necessarily exist an initialization $\varphi_1(\tau)$ of (1.5) such that $x(t - 1) = \varphi_0(t - 1)$ for $t \in [0, 1)$.

The trajectories of system (1.4), (1.5) are special trajectories generated by

$$\begin{pmatrix} \dot{z}_1(t) \\ \dot{z}_2(t) \end{pmatrix} = \begin{pmatrix} -z_2(t) \\ -v(t) \end{pmatrix}$$

as long as $z_2(t) = z_1(t - 1)$ and $v(t) = z_2(t - 1) = z_1(t - 2)$.

More generally, consider again the class of systems (1.2), (1.3). It is natural to rename the delayed state variables as $x(t - i) = z_i$, as well as the delayed control variable as $u(t - i) = v_i$. In this way, one gets the cascade system

$$\dot{z}_0 = F(z_0, \ldots z_s) + \sum_{i=0}^{l} G_i(z_0, \ldots z_s) v_i$$

$$\dot{z}_1 = F(z_1, \ldots z_{s+1}) + \sum_{i=0}^{l} G_i(z_1, \ldots z_{s+1}) v_{i+1} \tag{1.6}$$

$$\vdots$$

which has the following nice block representation given in Fig. 1.1 where Σ_0 is the system described by z_0, with entries $V_0 = (v_0, \ldots v_l)$ and $(z_1, \ldots, z_s)$; Σ_1 is the system described by the dynamics of z_1, with entries $V_1 = (v_1, \ldots v_{l+1}) = V_0(t - 1)$ and $(z_2, \ldots, z_{s+1})$; and Σ_2 is the system described by the dynamics of z_2, with entries $V_2 = (v_2, \ldots v_{l+2})$ and $(z_3, \ldots, z_{s+2}), \ldots$.

It is immediately understood, however, that the block scheme in Fig. 1.1 as well as the differential equations (1.6) represents a broader class of systems, not necessarily delayed and generated by Eq. (1.2). In fact, in order to represent the delay-time system (1.2), one cannot neglect that the variables $z_i(t)$ represent pieces of the same trajectory, which means that they cannot be initialized in an arbitrary way: for any real ℓ, and integer h such that $h \leq \ell \leq h + 1$, at any time t they have to satisfy the relation that $z_i(t - \ell) = z_{i+j}(t - \ell + j)$ for any integer $j \in [1, h]$. In a similar

Fig. 1.1 Block scheme of
system (1.2), (1.3)

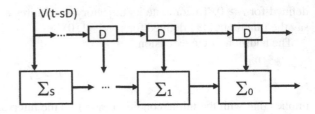

vein, the inputs $V_0, V_1, \ldots$ are generated by a unique signal u by considering also its
repeated delays. They are thus not independent.

Consider, for instance, the nonlinear time-delay system

$$\dot{x}(t) = \begin{pmatrix} x_1(t - \tau) \\ 1 \end{pmatrix} u(t) \tag{1.7}$$

with τ as a constant delay. As the single delayed variable is $x_1(t - \tau)$, one may
extend the dynamics (1.7) with

$$\dot{x}_1(t - \tau) = x_1(t - 2\tau)u(t - \tau)$$

and further by

$$\dot{x}_1(t - 2\tau) = x_1(t - 3\tau)u(t - 2\tau)$$

$$\vdots$$

Note that for some special cases as

$$\dot{x}(t) = \begin{pmatrix} x_2(t - \tau) \\ 1 \end{pmatrix} u(t), \tag{1.8}$$

it is sufficient to include the following extension:

$$\dot{x}_2(t - \tau) = u(t - \tau) \tag{1.9}$$

to describe the full behavior of the states.

Nevertheless, it will be seen later in this book that both systems (1.7) and (1.8)
are fully controllable. This is a major paradox with respect to delay-free driftless
systems. In fact, the underlying mathematical point is about the left-annihilator
$[1 \quad - x_1(t - \tau)]$ of the input matrix of system (1.7), or about the left-annihilator
$[1 \quad - x_2(t - \tau)]$ of the input matrix of system (1.8). These left-annihilators are com-
monly denoted as $dx_1(t) - x_1(t - \tau)dx_2(t)$ (respectively $dx_1(t) - x_2(t - \tau)dx_2(t)$).
Due to the delay, both one-forms $dx_1(t) - x_i(t - \tau)dx_2(t)$, for $i = 1, 2$, are not integrable and thus they are not representative of a noncontrollable state. More precisely,
this fact gives rise to issues on the characterization of integrability as discussed further
hereafter.

1.2 Integrability

A major difficulty in analyzing time-delay systems is their infinite dimensionality. Thus, in the nonlinear case, integrability results provided by Poincaré lemma or Frobenius theorem have to be revised.

Consider, for instance, the one form derived from system (1.8) for $\tau = 1$

$$\omega(x(t), x(t-1)) = dx_1 - x_2(t-1)dx_2.$$

It is not integrable, in the sense that there is no finite index j, no function $\lambda(x(t), x(t-1), \ldots, x(t-j))$ and no coefficient $\alpha(x(t), x(t-1), \ldots, x(t-j))$ such that $d\lambda(\cdot) = \alpha(\cdot)\omega(x(t), x(t-1))$.

The approach proposed in this book allows us to address successfully this kind of problems. This issue is analyzed in detail in Chap. 3.

1.3 Geometric Behavior

Consider again the nonlinear time-delay system (1.8)

$$\dot{x}(t) = \begin{pmatrix} x_2(t-\tau) \\ 1 \end{pmatrix} u(t).$$

In Fig. 1.2, the trajectory of the system is shown for a switching sequence of the input signal. The input switches from 1 to -1 includes five such forward and backward cycles. Differently from what would happen in the delay-free case when the input switches, the trajectory does not stay on the same integral manifold of one single vector field. A new direction is taken in the motion, which shows that the delay adds some additional freedom for the control direction and yields accessibility for the example under consideration. This is a surprising property of single-input driftless nonlinear time-delay systems and contradicts pre-conceived ideas as it could not happen for delay-free systems. As it will be argued in Sect. 2.5, the motion in the x_1 direction of the final point of each cycle has to be interpreted as the motion along the nonzero Lie Bracket of the delayed control vector field with itself. For instance, system (1.8) with its extension (1.9) reads

$$\dot{x}(t) = \begin{pmatrix} x_2(t-\tau) \\ 1 \\ 0 \end{pmatrix} u(t) + \begin{pmatrix} 0 \\ 0 \\ 1 \end{pmatrix} u(t-\tau).$$

The Lie Bracket of the two vector fields generates a third independent direction. These general intuitive considerations are discussed formally in the book using precise definitions.

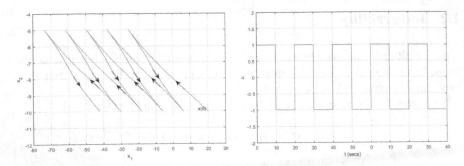

Fig. 1.2 Forward and backward integration yields a motion in an additional specific direction. On the left the state trajectory of the system initialized with $x_2 = -10$ for $t \in [-2, 0)$. The delay is fixed to $\tau = 2\,\text{secs}$

1.4 Accessibility and Observability Properties

The accessibility/controllability of systems (1.7) and (1.8) has already been discussed above in relation with the topic of integrability. This property is unexpected for a driftless single-input system.

Now, consider, for instance,

$$\dot{x}(t) = \begin{pmatrix} x_2(t - \tau) \\ x_3(t) \\ 1 \end{pmatrix} u(t). \tag{1.10}$$

System (1.10) is not fully controllable, as for $\tau \neq 0$ there is one autonomous element $\lambda = x_3^2/2 - x_2$.[1] Define the change of variables $(z_1(t), z_2(t), z_3(t)) = (x_1(t), x_3(t), x_3^2(t)/2 - x_2(t))$. The dynamics in the new system of coordinates decomposes system (1.10) into the fully controllable subsystem

$$\begin{pmatrix} \dot{z}_1(t) \\ \dot{z}_2(t) \end{pmatrix} = \begin{pmatrix} z_2^2(t - \tau)/2 - z_3(t - \tau) \\ 1 \end{pmatrix} u(t)$$

and the noncontrollable subsystem

$$\dot{z}_3(t) = 0.$$

More generally, it will be shown that a decomposition with respect to accessibility can always be carried out and a full characterization of accessibility can be given in terms of a rank condition for nonlinear time-delay systems as shown in Chap. 4. This result is in the continuation of the celebrated geometric approach for delay-free systems;

---

[1]For $\tau = 0$ one would reduce to the delay-free case with two autonomous elements, $\lambda_1 = x_3^2/2 - x_2$ and $\lambda_2 = -x_3^3/3 + x_2 x_3 - x_1$.

the work (Hermann 1963) on accessibility has certainly been "the seminal paper inspiring the geometric approach that started to be developed by Lobry, Jurdjevic, Sussmann, Hermes, Krener, Sontag, Brockett in the early 1970's" (quoted from Sallet 2007).

The above discussion enlightens also that in the delay context two different notions of accessibility should be considered and accordingly two different criteria. A first notion, which could be considered as an immediate generalization of Kalman's result, is based on the consideration of autonomous elements for the given system. Such autonomous elements may be functions of the current state and its delays. Looking instead at accessibility as the possibility to define a control which allows to move from some initial point x_0 at time t_0 to a given final point x_f at some time t leads to a different notion and characterization of accessibility, actually a peculiarity of time-delay systems only. This new notion will be referred to t-accessibility. The difference between the two cases is illustrated by the following simple linear example.

Example 1.2 Consider the linear system

$$\dot{x}_1(t) = u(t)$$
$$\dot{x}_2(t) = u(t-1).$$

Such a system is not fully accessible due to the existence of the autonomous element $\lambda = x_2(t) - x_1(t-1)$. This means that the point reachable at time t is linked to the point reached at time $t-1$ through the relation $x_2(t) - x_1(t-1) = constant$. However, we may reach a given fixed point at a different time $\bar{t}$.

Assume, for instance, that $x_1(t) = x_{10}, x_2(t) = x_{20}, u(t) = 0$ for $t \in [-1, 0)$. And let $x_f = (x_{1f}, x_{2f})^T$ be the final point to be reached. Then on the interval $[0, 1)$

$$\dot{x}_1(t) = u(t)$$
$$\dot{x}_2(t) = 0,$$

so that $x_2(t) = x_{20}$ and for a constant value of the control $u(t) = u_0$, $x_1(t) = u_0 t + x_{10}$ for $t \in [0, 1)$. If $x_{2f} \neq x_{20}$ it is immediately clear that one cannot reach the given final point for any $t \in [0, 1)$. Now let the control change to $u(t) = u_1$ on the interval $t \in [1, 2)$. Then on such interval

$$\dot{x}_1(t) = u_1$$
$$\dot{x}_2(t) = u_0$$

and $x_1(t) = (t-1)u_1 + x_1(1) = (t-1)u_1 + u_0 + x_{10} = x_{1f}$ and $x_2(t) = (t-1)u_0 + x_{20} = x_{2f}$, so that once $t > 1$ is fixed, one gets

$$u_0 = \frac{x_{2f} - x_{20}}{t-1}$$

$$u_1 = \frac{x_{1f} - x_{10} - u_0}{t-1}.$$

The situation is quite different when considering the property of observability. For instance, consider the single-output time-delay system

$$\dot{x}_1(t) = 0$$
$$\dot{x}_2(t) = 0$$
$$y(t) = x_1(t)x_1(t - \tau) + x_2(t)x_2(t - \tau).$$

As any time derivative of the output is zero, for $t \geq 0$, the two state variables of the above system cannot be estimated independently and the system is not fully observable. As a matter of fact, differently from the delay-free case, there is no *invertible* change of state coordinates which decomposes the system into an observable subsystem and a nonobservable one. This contradicts common beliefs on this matter. Additional assumptions are required (Zheng et al. 2011) to ensure that such a decomposition still exists. This topic is addressed in Chap. 5.

1.5   Notation

This paragraph is devoted to the notation which will be used in the book. In general, we will refer to the class of multi-input multi-output nonlinear time-delay systems

$$\dot{x}(t) = F(x(t), \ldots, x(t - sD)) + \sum_{i=0}^{l} \sum_{j=1}^{m} G_{ji}(x(t), \ldots, x(t - sD))u_j(t - iD))$$
$$y(t) = H(x(t), \ldots, x(t - sD)),$$

(1.11)

where $x(t) \in I\!R^n$ and $u(t) \in I\!R^m$ are the current values of the state and control variables; D is a constant delay; $s, l \geq 0$ are integers; and the functions $G_{ji}(x(t), \ldots, x(t - sD))$, $j \in [1, m]$, $i \in [0, l]$, $F(x(t), \ldots, x(t - sD))$, and $H(x(t), \ldots, x(t - sD))$ are analytic in their arguments. It is easy to see that such a class of systems includes the case of constant multiple commensurate delays as well (Gu et al. 2003).

The following notation taken from Califano and Moog (2017), Xia et al. (2002) will be extensively used:

- $\mathbf{x}_{[p,s]}^T = (x^T(t + pD), \ldots x^T(t - sD)) \in I\!R^{(p+s+1)n}$, denotes the vector consisting of the np *future* values $x(t + iD)$, $i \in [1, p]$, of the state together with the first $(s + 1)n$ components of the state of the infinite dimensional system associated to (1.11). When $p = 0$, the more simple notation $\mathbf{x}_{[s]}^T = \mathbf{x}_{[0,s]}^T \in I\!R^{(s+1)n}$ is used, with $\mathbf{x}_{[0]} = [x_{1,[0]}, \ldots, x_{n,[0]}]^T = x(t) \in I\!R^n$, $\mathbf{u}_{[0]} = [u_{1,[0]}, \ldots, u_{m,[0]}]^T = u(t) \in I\!R^m$, the current values of the state and input variables.
- $\mathbf{x}_{[p,s]}^T(-i) = (x^T(t + pD - iD), \ldots x^T(t - sD - iD))$. Accordingly, $\mathbf{x}_{[s]}(-i) = \mathbf{x}_{[0,s]}(-i)$; $x_{j,[0]}(-i) := x_j(t - iD)$, and $u_{\ell,[0]}(-i) := u_\ell(t - iD)$ denote, respectively, the j-th and ℓ-th components of the instantaneous values of the state and

input variables delayed by $\tau = iD$. When no confusion is possible the subscript
will be omitted so that $\mathbf{x}$ will stand for $\mathbf{x}_{[p,s]}$, while $\mathbf{x}(-i)$ will stand for $\mathbf{x}_{[p,s]}(-i)$.

- $\mathbf{u}^{[j]} := (\mathbf{u}^T, \dot{\mathbf{u}}^T, \ldots, (\mathbf{u}^{(j)})^T)^T$ where $\mathbf{u}^{[-1]} = \emptyset$.
- $\mathcal{K}^*$ denotes the field of meromorphic functions $f(\mathbf{x}_{[p,s]}, \mathbf{u}^{[k]}_{[q,j]})$, with p, s, k, q, $j \in \mathbb{N}$. The subfield $\mathcal{K}$ of $\mathcal{K}^*$, consisting of causal meromorphic functions, is obtained for $p = q = 0$.
- Given a function $f(\mathbf{x}_{[p,s]}, \mathbf{u}^{[k]}_{[q,j]})$, the following notation is in order:

$$f(-l) := f(\mathbf{x}_{[p,s]}(-l), \mathbf{u}^{[k]}_{[q,j]}(-l)).$$

As an example, the function $f(x) = x_1(t)x_2(t - D) - x_1^2(t - 2D)$ in the previous notation is written as $f(\mathbf{x}) = x_{1,[0]}x_{2,[0]}(-1) - x_{1,[0]}^2(-2)$ and accordingly

$$f(-5) = x_{1,[0]}(-5)x_{2,[0]}(-6) - x_{1,[0]}^2(-7).$$

- d is the standard differential operator.
- δ represents the backward time-shift operator: for $a(\cdot), f(\cdot) \in \mathcal{K}^*$: $\delta[a\,df] = a(-1)\delta df = a(-1)df(-1)$.
- Let for $i \in [1, j]$, $\tau_i(\mathbf{x}_{[l]})$ be vector fields defined in an open set $\Omega_l \subseteq \mathbb{R}^{n(l+1)}$. Then $\Delta = \text{span}\{\tau_i(\mathbf{x}_{[l]}), i = 1, \ldots, j\}$ represents the distribution generated by the vector fields $\tau_i(\cdot)$ and defined on $\mathbb{R}^{n(l+1)}$. $\bar{\Delta}$ represents its involutive closure, that is, for any two vector fields $\tau_i(\cdot)$, $\tau_j(\cdot) \in \bar{\Delta}$ the Lie Bracket $[\tau_i, \tau_j] = \frac{\partial \tau_i}{\partial \mathbf{x}_{[l]}}\tau_j - \frac{\partial \tau_j}{\partial \mathbf{x}_{[l]}}\tau_i \in \bar{\Delta}$ (Isidori 1995).
  $\Delta_{[p,q]}$ will denote a distribution in $\text{span}_{\mathcal{K}^*}\{\frac{\partial}{\partial \mathbf{x}_{[0]}(p)}, \ldots, \frac{\partial}{\partial \mathbf{x}_{[0]}(-q)}\}$.
- Let $\mathcal{E}$ denote the vector space spanned by the differentials $\{dx(t - i); i \in \mathbb{Z}\}$ over the field $\mathcal{K}^*$. The elements of $\mathcal{E}$ are called one-forms. A one-form ω is said to be exact if there exists $\varphi \in \mathcal{K}^*$ such that $\omega = d\varphi$. The use of exterior differentiation and of the wedge product allows to state in a concise manner both Poincaré lemma and Frobenius theorem (Choquet-Bruhat et al. 1989):

  – The one-form ω is locally exact if and only if $d\omega = 0$.
  – The codistribution $\text{span}_{\mathcal{K}}\{\omega_1, \ldots, \omega_q\}$ is integrable if and only if the $(q + 2)$-forms $d\omega_i \wedge \omega_1 \wedge \cdots \wedge \omega_q$ are zero for $i = 1, \ldots, q$, where $\wedge$ denotes the wedge product of differential forms (Choquet-Bruhat et al. 1989).
  – The following notation is also used:

$$d\omega = 0 \quad \text{mod} \ \text{span}_{\mathcal{K}}\{\bar{\omega}_1, \ldots, \bar{\omega}_q\},$$

which means that $d\omega \wedge \bar{\omega}_1 \wedge \cdots \wedge \bar{\omega}_q = 0$.

With such a notation the given system (1.1) is rewritten as

$$\Sigma : \begin{cases} \dot{\mathbf{x}}_{[0]} = F(\mathbf{x}_{[s]}) + \sum_{j=1}^{m}\sum_{i=0}^{l} G_{ji}(\mathbf{x}_{[s]})\mathbf{u}_{[0],j}(-i) \\ \mathbf{y}_{[0]} = H(\mathbf{x}_{[s]}) \end{cases} \tag{1.12}$$

1.6 Recalls on Non-commutative Algebra

Non-commutative algebra is used throughout the book to address the study of time-delay systems. In this section, the mathematics and definitions beyond this method are introduced (see, for example, Cohn 1985, Banks 2002).

Let us now consider the time-shift operator δ recalled in the previous paragraph, and let us denote by $\mathcal{K}^*(\delta]$ the (left-) ring of polynomials in δ with coefficients in $\mathcal{K}^*$ (analogously $\mathcal{K}(\delta]$ will denote the (left-) ring of polynomials in δ with coefficients in $\mathcal{K}$). Every element of $\mathcal{K}^*(\delta]$ may be written as

$$\alpha(\delta] = \alpha_0(\mathbf{x}) + \alpha_1(\mathbf{x})\delta + \cdots + \alpha_{r_\alpha}(\mathbf{x})\delta^{r_\alpha}, \qquad \alpha_j(\cdot) \in \mathcal{K}^*$$

$r_\alpha = \deg(\alpha(\delta])$ is the polynomial degree of $\alpha(\delta]$ in δ. Let

$$\beta(\delta] = \beta_0(\mathbf{x}) + \beta_1(\mathbf{x})\delta + \cdots + \beta_{r_\beta}(\mathbf{x})\delta^{r_\beta}$$

be another element of $\mathcal{K}^*(\delta]$ of polynomial degree r_β. Then addition and multiplication on this ring are defined by Xia et al. (2002):

$$\alpha(\delta] + \beta(\delta] = \sum_{i=0}^{\max\{r_\alpha, r_\beta\}} (\alpha_i + \beta_i)\delta^i$$

and

$$\alpha(\delta]\beta(\delta] = \sum_{i=0}^{r_\alpha}\sum_{j=0}^{r_\beta} \alpha_i \, \beta_j(-i)\delta^{i+j}.$$

Analogously, we can consider vectors, covectors, and matrices whose entries are in the ring. The standard operations of sum and product are well defined once one applies the previous rules on the sum and product of the elements of the ring. As for matrices, it should be noted that in this case the property of full rank of a square matrix does not imply automatically the existence of its inverse. If the inverse exists, a stronger property is satisfied which is that of unimodularity of the matrix. We give hereafter the formal definition of unimodular matrix which will play a fundamental role in the definitions of changes of coordinates. Some examples will clarify the difference with full rank but not unimodular matrices.

Definition 1.1 (Cohn 1985) A matrix $A(\delta) \in \mathcal{K}^*(\delta]^{k \times k}$ is unimodular if it is invertible within the ring of polynomial matrices, i.e. if there exists a $B(\delta) \in \mathcal{K}^*(\delta]^{k \times k}$ such that $A(\delta)B(\delta) = B(\delta)A(\delta) = I$.

Example 1.3 The matrix

$$A(\delta) = \begin{pmatrix} 1 & x(t-1)\delta \\ \delta & 1 + x(t-2)\delta^2 \end{pmatrix}$$

is unimodular, since it admits an inverse which is given by

$$A^{-1}(\delta) = \begin{pmatrix} 1 + x(t-1)\delta^2 & -x(t-1)\delta \\ -\delta & 1 \end{pmatrix}.$$

In fact $A(\delta)A^{-1}(\delta) = A^{-1}(\delta)A(\delta) = I$. Note that while any unimodular matrix has full rank, the converse is not true. For example, there is no polynomial inverse for $A(\delta) = (1 + \delta)$.

Let us now consider a set of one-forms. It is immediately clear that such a set has both the structure of a vector space $\mathcal{E}$ over the field $\mathcal{K}^*$ and the structure of a module, denoted $\mathcal{M}$, over the ring $\mathcal{K}^*(\delta]$, i.e.

$$\mathcal{M} = \mathrm{span}_{\mathcal{K}^*(\delta]}\{dx(t)\}.$$

As an example, the one-form $\omega = x(t)dx(t-1) - x^2(t-1)dx(t)$ can also be rewritten as $\omega = \left(\mathbf{x}_{[0]}\delta - x^2(-1)\right)d\mathbf{x}_{[0]}$. This twofold possibility allows us to interpret the given one forms in two different ways, as shown in the next example.

Example 1.4 The one-forms $dx_1(t)$ and $dx_1(t-1)$ are independent over the field $\mathcal{K}$, while they are dependent over the ring $\mathcal{K}(\delta]$, since $\delta dx_1(t) - dx_1(t-1) = 0$. This simple example shows that the action of time delay is taken into account in $\mathcal{M}$, but not in $\mathcal{E}$. This motivates the definition of the module $\mathcal{M}$.

A left-submodule of $\mathcal{M}$ consists of all possible linear combinations of given one-forms (or row vectors) $\{\omega_1, \ldots, \omega_k\}$ over the ring $\mathcal{K}^*(\delta]$, i.e. linear combinations of row vectors. A left-submodule, generated by $\{\omega_1, \ldots, \omega_j\}$, is denoted by $\Omega = \mathrm{span}_{\mathcal{K}^*(\delta]}\{\omega_1, \ldots, \omega_j\}$.

Any $\omega(\mathbf{x}, \delta)d\mathbf{x}_{[0]} \in \Omega(\delta]$ can be expressed as a linear combination of the generators of $\Omega(\delta]$, that is,

$$\omega(\mathbf{x}, \delta)d\mathbf{x}_{[0]} = \sum_{i=1}^{j} \alpha_i(\mathbf{x}, \delta)\omega_i(\mathbf{x}, \delta)d\mathbf{x}_{[0]}.$$

An important property of the considered modules is given by the so-called closure, introduced in Conte and Perdon (1995). Before giving the formal definition we will illustrate it through an example.

Example 1.5 Consider the left-submodule $\Omega(\delta] = \mathrm{span}_{\mathcal{K}^*(\delta]}\{dx_1(t-1), dx_2(t)\}$. Clearly the rank of $\Omega(\delta]$ is 2. Any $\omega \in \Omega(\delta] = \alpha_1(\mathbf{x}, \delta)dx_1(t-1) + \alpha_2(\mathbf{x}, \delta)dx_2(t)$. However, there are some $\bar{\omega}$ such that $\bar{\omega} \notin \Omega(\delta]$, while $\alpha_0(\mathbf{x}, \delta)\bar{\omega} \in \Omega(\delta]$. This is, for example, the case of $\bar{\omega} = dx_1(t)$; we will then say that $\Omega(\delta]$ is not left-closed.

We thus can give the following definition from Conte and Perdon (1995).

Definition 1.2 Let $\Omega(\delta] = \text{span}_{\mathcal{K}^*(\delta]}\{\omega_1(\mathbf{x}, \delta)d\mathbf{x}_{[0]}, \ldots \omega_j(\mathbf{x}, \delta)d\mathbf{x}_{[0]}\}$ be a left-submodule of rank j with $\omega_i \in \mathcal{K}^{*(1 \times n)}(\delta]$. The left closure of $\Omega(\delta]$ is the largest left-submodule $\Omega_c(\delta]$ of rank j containing $\Omega(\delta]$.

The left closure of the left-submodule Ω is thus the largest left-submodule, containing Ω, with the same rank as Ω.

Similar conclusions can be obtained on right-submodules. More precisely, a right-module of $\hat{\mathcal{M}}$ (Califano et al. 2011a) consists of all possible linear combinations of column vectors $\tau_1, \ldots, \tau_j$, $\tau_i \in \mathcal{K}^{n \times 1}(\delta]$, and is denoted by $\Delta = \text{span}_{\mathcal{K}^*(\delta]}\{\tau_1, \ldots, \tau_j\}$.

Let $\Delta(\delta] = \text{span}_{\mathcal{K}^*(\delta]}\{\tau_1(\mathbf{x}, \delta), \ldots \tau_j(\mathbf{x}, \delta)\}$ be a right-submodule of rank j with $\tau_i \in \mathcal{K}^{*(n \times 1)}(\delta]$. Then any $\tau(\mathbf{x}, \delta) \in \Delta(\delta]$ can be expressed as $\tau(\mathbf{x}, \delta) = \sum_{i=1}^{j} \tau_i(\mathbf{x}, \delta)\alpha_i(\mathbf{x}, \delta)$.

Accordingly, the following definition of the right closure of a right-submodule can be given.

Definition 1.3 Let $\Delta(\delta] = \text{span}_{\mathcal{K}^*(\delta]}\{\tau_1(\mathbf{x}, \delta), \ldots \tau_j(\mathbf{x}, \delta)\}$ be a right-submodule of rank j with $\tau_i \in \mathcal{K}^{*(n \times 1)}(\delta]$. The right closure of $\Delta(\delta]$ is the largest right-submodule $\Delta_c(\delta]$ of rank j containing $\Delta(\delta]$.

Definition 1.4 The right closure of a right-submodule Δ of $\hat{\mathcal{M}}$, denoted by $cl_{\mathcal{K}(\vartheta]}(\Delta)$, is defined as $cl_{\mathcal{K}(\vartheta]}(\Delta) = \{X \in \hat{\mathcal{M}} \mid \exists q(\vartheta) \in \mathcal{K}^*(\vartheta], Xq(\vartheta) \in \Delta\}$.

The right closure of the right-submodule Δ is the largest right-submodule, containing Δ, with the same rank as Δ.

Of course, one may consider the right-annihilator of a left-submodule or the left-annihilator of a right-submodule. In both cases, it is easily seen that one ends up on a closed submodule. In fact

Definition 1.5 The right-kernel (right-annihilator) of the left-submodule Ω is the right-submodule Δ containing all vectors $\tau(\delta) \in \hat{\mathcal{M}}$ such that $P(\delta)\tau(\delta) = 0$.

And it is easily verified that by definition the right-kernel is necessarily closed. Analogously

Definition 1.6 The left-kernel (left-annihilator) of Δ is the left-submodule Ω containing all one forms $\omega(\delta) \in \mathcal{M}$ such that $\omega(\delta)\Delta = 0$.

Again, by definition, the left-kernel is necessarily closed.

An immediate consequence is that the right-kernel of a left-submodule and of its closure coincide. Analogously, the left-kernel of a right-submodule and of its closure coincide.

We end the chapter by recalling the relations between the degrees of a submodule and its left-annihilator which were shown in Califano and Moog (2017).

Lemma 1.1 *Consider the matrix*

$$\Gamma(\mathbf{x}_{[p,s]}, \delta) = (\tau_1(\mathbf{x}_{[p,s]}, \delta), \ldots, \tau_j(\mathbf{x}_{[p,s]}, \delta)).$$

Let $\bar{s} = deg(\Gamma(\mathbf{x}_{[p,s]}, \delta))$. The left-annihilator $\Omega(\mathbf{x}_{[\bar{p},\alpha]}, \delta)$ satisfies the following relations:

(i) $deg(\Omega(\mathbf{x}_{[\bar{p},\alpha]}, \delta)) \leq j \, [deg(\Gamma(\mathbf{x}, \delta))];$
(ii) $\bar{p}, \alpha$ can be chosen to be $\alpha \leq s + deg(\Omega(\mathbf{x}, \delta)), \bar{p} \leq p$.

Consequently, if $\Gamma(\mathbf{x}, \delta)$ is causal, then $\Omega(\mathbf{x}, \delta)$ is also causal.

Proof Without loss of generality, assume that the first j rows of $\Gamma(\mathbf{x}_{[p,s]}, \delta)$ are linearly independent over $\mathcal{K}(\delta)$. Then $\Omega(\mathbf{x}_{[\bar{p},\alpha]}, \delta)$ must satisfy

$$\Omega(\mathbf{x}, \delta)\Gamma(\mathbf{x}, \delta) = [\Omega_1(\mathbf{x}, \delta), \Omega_2(\mathbf{x}, \delta)] \begin{pmatrix} \Gamma_1(\mathbf{x}, \delta) \\ \Gamma_2(\mathbf{x}, \delta) \end{pmatrix} = 0,$$

where $\Gamma_1(\mathbf{x}, \delta)$ is a $j \times j$ full rank matrix, accordingly $\Gamma_2(\mathbf{x}, \delta)$ is a $(n - j) \times j$ matrix, $\Omega_1(\mathbf{x}, \delta)$ is a $(n - j) \times j$ matrix, and $\Omega_2(\mathbf{x}, \delta)$ is a $(n - j) \times (n - j)$ matrix. Let $r_{\Omega_1} = deg(\Omega_1(\mathbf{x}, \delta))$, $r_{\Omega_2} = deg(\Omega_2(\mathbf{x}, \delta))$, $r_{\Gamma_1} = deg(\Gamma_1(\mathbf{x}, \delta))$, and $r_{\Gamma_2} = deg(\Gamma_2(\mathbf{x}, \delta))$. Then we have that $r_{\Omega_1} + r_{\Gamma_1} = r_{\Omega_2} + r_{\Gamma_2}$. $\Omega(\mathbf{x}, \delta)$ must satisfy the following relations:

$$\Omega_1^0 \Gamma_1^0 + \Omega_2^0 \Gamma_2^0 = 0$$
$$\Omega_1^0 \Gamma_1^1 + \Omega_1^1 \Gamma_1^0(-1) + \Omega_2^0 \Gamma_2^1 + \Omega_2^1 \Gamma_2^0(-1) = 0$$

$$\vdots \tag{1.13}$$

$$\sum_{i=1}^{2} \sum_{j=0}^{r_{\Omega_i}} \Omega_i^j \Gamma_i^{r_{\Omega_i}+r_{\Gamma_i}-j}(-j) = 0.$$

We have $(n - j)j(r_{\Omega_1} + 1)$ unknowns for Ω_1, $(n - j)^2(r_{\Omega_2} + 1)$ unknowns for Ω_2 and $(n - j)j(r_{\Omega_1} + r_{\Gamma_1} + 1)$ equations. In order to be sure to get a solution

$$(n - j)j(r_{\Omega_1} + 1) + (n - j)^2(r_{\Omega_2} + 1) \geq (n - j)j(r_{\Omega_1} + r_{\Gamma_1} + 1),$$

that is, $(n - j)(r_{\Omega_2} + 1) \geq jr_{\Gamma_1}$. Once fixed r_{Ω_2}, we get that $r_{\Omega_1} = r_{\Omega_2} + r_{\Gamma_2} - r_{\Gamma_1}$. In the worst case, $r_{\Gamma_2} = r_{\Gamma_1} = r$ and $n - j = 1$, which proves (i).

From the set of equations (1.13), fixing the independent parameters as functions of $\mathbf{x}_{[0]}$ only, then the maximum delay is given by the largest between $(s + r_{\Omega_1}, s + r_{\Omega_2})$, while $\bar{p} \leq p$ which proves (ii). Consequently, if $\Gamma(\mathbf{x}, \delta)$ is causal, then $p = 0$ so that $\bar{p} \leq 0$ which shows that $\Omega(\mathbf{x}, \delta)$ is also causal.

Chapter 2
Geometric Tools for Time-Delay Systems

In this chapter, we introduce the main tools that will be used in the book to deal with nonlinear time-delay systems affected by constant commensurate delays. We will introduce basic notions such as the Extended Lie derivative and the Polynomial Lie Bracket (Califano et al. 2011a; Califano and Moog 2017) which generalize to the time-delay context the standard definitions of Lie derivative and Lie Bracket used to deal with nonlinear systems. We will finally show how changes of coordinates and feedback laws act on the class of systems considered.

Before going into the technical details let us first recall that with the notation introduced in Chap. 1, we can rewrite system (1.1) as

$$\dot{\mathbf{x}}_{[0]} = F(\mathbf{x}_{[s]}) + \sum_{i=0}^{l} \sum_{j=1}^{m} G_{ji}(\mathbf{x}_{[s]}) u_{j,[0]}(-i) \tag{2.1}$$

$$y_{j,[0]} = H_j(\mathbf{x}_{[s]}), \qquad j \in [1, p]. \tag{2.2}$$

As it was already discussed, the following infinite dimensional dynamics is naturally associated to the dynamics (2.1):

$$\dot{\mathbf{x}}_{[0]} = F(\mathbf{x}_{[s]}) + \sum_{i=0}^{l} \sum_{j=1}^{m} G_{ji}(\mathbf{x}_{[s]}) u_{j,[0]}(-i)$$

$$\dot{\mathbf{x}}_{[0]}(-1) = F(\mathbf{x}_{[s]}(-1)) + \sum_{i=0}^{l} \sum_{j=1}^{m} G_{ji}(\mathbf{x}_{[s]}(-1)) u_{j,[0]}(-i-1) \tag{2.3}$$

$$\vdots$$

C. Califano and C. H. Moog, *Nonlinear Time-Delay Systems*,
SpringerBriefs in Control, Automation and Robotics,
https://doi.org/10.1007/978-3-030-72026-1_2

The advantage in the representation (2.3) is that it shows that the given time-delay system can be represented as an interconnection of subsystems and that these subsystems are coupled through the action of the control. Nevertheless caution must be used when referring to this last representation since the variables $x(-i)$ are connected to each other through time. This will be further discussed in this chapter.

2.1 The Initialization of the Time-Delay System Versus the Initialization of the Delay-Free Extended System

Consider a truncation of system (2.3):

$$\dot{\mathbf{x}}_{[0]} = F(\mathbf{x}_{[s]}) + \sum_{i=0}^{l} \sum_{j=1}^{m} G_{ji}(\mathbf{x}_{[s]}) u_{j,[0]}$$

$$\dot{\mathbf{x}}_{[0]}(-1) = F(\mathbf{x}_{[s]}(-1)) + \sum_{i=0}^{l} \sum_{j=1}^{m} G_{ji}(\mathbf{x}_{[s]}(-1)) u_{j,[0]}(-i-1) \qquad (2.4)$$

$$\vdots$$

$$\dot{\mathbf{x}}_{[0]}(-k) = F(\mathbf{x}_{[s]}(-k)) + \sum_{i=0}^{l} \sum_{j=1}^{m} G_{ji}(\mathbf{x}_{[s]}(-k)) u_{j,[0]}(-i-k).$$

Rename $v_{j0} = u_{j,[0]}$, $v_{j\ell} = u_{j,[0]}(-\ell)$ for $\ell = 1, \ldots, l+k$, so that system (2.4) reads

$$\dot{\mathbf{x}}_{[0]} = F(\mathbf{x}_{[s]}) + \sum_{i=0}^{l} \sum_{j=1}^{m} G_{ji}(\mathbf{x}_{[s]}) v_{j,i}$$

$$\dot{\mathbf{x}}_{[0]}(-1) = F(\mathbf{x}_{[s]}(-1)) + \sum_{i=0}^{l} \sum_{j=1}^{m} G_{ji}(\mathbf{x}_{[s]}(-1)) v_{j,i+1} \qquad (2.5)$$

$$\vdots$$

$$\dot{\mathbf{x}}_{[0]}(-k) = F(\mathbf{x}_{[s]}(-k)) + \sum_{i=0}^{l} \sum_{j=1}^{m} G_{ji}(\mathbf{x}_{[s]}(-k)) v_{j,i+k}$$

$$\dot{\mathbf{x}}_{[0]}(-k-1) = \phi_1$$

$$\vdots$$

$$\dot{\mathbf{x}}_{[0]}(-k-s) = \phi_s.$$

While system (2.1) requires an initial condition on the time interval $[-s, 0)$ to compute its trajectories for $t \geq 0$, system (2.5) requires an initialization on the time interval $[-(k+s), 0)$. As a consequence, any trajectory of (2.5) is also a trajectory of (2.1), provided the latter is correctly initialized on the interval $[-s, 0)$.

On the contrary, considering a given trajectory of system (2.1), there may not necessarily exist a corresponding initialization function defined on the time interval $[-(k+s), 0)$ for system (2.5) which reproduces the given trajectory of system (2.1). This is a direct consequence of the fact that the initialization of the system is not necessarily a solution of the set of differential equations. This point is further clarified through the next example.

Example 2.1 Consider the dynamics

$$\dot{x} = -x(t-1) \tag{2.6}$$

with $x(\tau) = 1$ for $\tau \in [-1, 0)$. System (2.6) is extended to

$$\begin{aligned} \dot{x}(t) &= -x(t-1) \\ \dot{x}(t-1) &= -x(t-2). \end{aligned} \tag{2.7}$$

There is no initialization function for system (2.7) over the time interval $[-2, -1)$ so that the trajectory of (2.7) coincides with the trajectory of (2.6) for any $t \geq -1$. In the special example, an ideal Dirac impulse at time $t = -1$ is required to achieve the reproduction of the trajectory.

Coming back to the extended system (2.5), note that the further trick which consists in renaming $x_0 = x_{[0]}$, $x_\ell = x_{[0]}(-\ell)$ for $\ell = 1, \ldots, s+k$ may be misleading as $x_0, \ldots, x_{s+k}$ are not independent.

$$\dot{x}_0 = F(x_0, \ldots, x_s) + \sum_{i=0}^{l} \sum_{j=1}^{m} G_{ji}(x_0, \ldots, x_s) v_{j,i}$$

$$\dot{x}_1 = F(x_1, \ldots, x_{s+1}) + \sum_{i=0}^{l} \sum_{j=1}^{m} G_{ji}(x_1, \ldots, x_{s+1}) v_{j,i+1} \tag{2.8}$$

$$\vdots$$

$$\dot{x}_k = F(x_k, \ldots, x_{s+k}) + \sum_{i=0}^{l} \sum_{j=1}^{m} G_{ji}(x_k, \ldots, x_{s+k}) v_{j,i+k}$$

$$\dot{x}_{k+1} = \phi_1$$

$$\vdots$$

$$\dot{x}_{k+s} = \phi_s.$$

From a practical point of view, given a nonlinear time-delay system, its representation (2.8) can be used to compute the solution of the system by referring to the so-called step method as shown in the example hereafter.

Example 2.2 Consider the dynamics

$$
\begin{aligned}
\dot{\mathbf{x}}(t) &= f(\mathbf{x}(t), \mathbf{x}(t-1)), \quad t \geq 0 \\
\mathbf{x}(t) &= \vartheta_0(t), \qquad -1 \leq t < 0.
\end{aligned}
\tag{2.9}
$$

To compute the solution starting from the initialization $\mathbf{x}(t) = \vartheta_0(t)$, for $-1 \leq t < 0$, the following steps are taken according to Garcia-Ramirez et al. (2016a):

- The solution $\vartheta_1(t)$ of (2.9) on the time interval $0 \leq t < 1$ is found as the solution of the delay-free system (2.8) subject to the appropriate initial condition $\vartheta_0(t)$:

$$
\begin{aligned}
\dot{\mathbf{x}}(t) &= f(\mathbf{x}(t), \vartheta_0(t-1)), \quad 0 \leq t < 1 \\
\mathbf{x}(t) &= \vartheta_0(t), \qquad -1 \leq t < 0.
\end{aligned}
\tag{2.10}
$$

- The solution $\vartheta_2(t)$ of (2.9) on the time interval $1 \leq t < 2$ is found as the solution of the delay-free system (2.8) subject to the initial condition $\vartheta_1(t)$ computed at the previous step:

$$
\begin{aligned}
\dot{\mathbf{x}}(t) &= f(\mathbf{x}(t), \vartheta_1(t-1)), \quad 1 \leq t < 2 \\
\mathbf{x}(t) &= \vartheta_1(t), \qquad 0 \leq t < 1.
\end{aligned}
\tag{2.11}
$$

- More generally, for $k \geq 0$, the solution $\vartheta_{k+1}(t)$ of (2.9) on the time interval $k \leq t < k+1$ is found as the solution of the delay-free system (2.8) subject to the initial condition $\vartheta_k(t)$:

$$
\begin{aligned}
\dot{\mathbf{x}}(t) &= f(\mathbf{x}(t), \vartheta_k(t-1)), \quad k \leq t < k+1 \\
\mathbf{x}(t) &= \vartheta_k(t), \qquad k-1 \leq t < k.
\end{aligned}
\tag{2.12}
$$

Note that shifting (2.10) into the interval $1 \leq t < 2$ yields

$$
\dot{\mathbf{x}}(t-1) = f(\mathbf{x}(t-1), \vartheta_0(t-2)), \quad 1 \leq t < 2.
$$

Thus, the set of equations (2.5) defined on the interval $(k-1) \leq t < k$ becomes in the case of this example

$$
\begin{aligned}
\dot{\mathbf{x}}(t) &= f(\mathbf{x}(t), \vartheta_{k-1}(t-1)), \\
\dot{\mathbf{x}}(t-1) &= f(\mathbf{x}(t-1), \vartheta_{k-2}(t-2)), \\
&\ \ \vdots \\
\dot{\mathbf{x}}(t-(k-1)) &= f(\mathbf{x}(t-(k-1)), \vartheta_0(t-k)).
\end{aligned}
\tag{2.13}
$$

System (2.13) defines the solution on the first k units of time of Eq. (2.9) shifted into the interval $(k-1) \leq t < k$, with initialization $x(\tau) = \vartheta_0(\tau)$ for $\tau \in [k-1, k)$.

Moreover, through a change of variable $x_0(t) = x(t)$, $x_1(t) = x(t-1)$, $x_k(t) = x(t-k)$, system (2.13) can also be represented as (2.8)

$$\dot{x}_0(t) = f(x_0(t), x_1(t)),$$
$$\dot{x}_1(t) = f(x_1(t), x_2(t)),$$
$$\vdots$$
$$\dot{x}_{k-1}(t) = f(x_{k-1}(t), \vartheta_0(t-1)),$$

with initial conditions $x_0(0) = x_1(1)$, $x_1(0) = x_2(1)$, $x_2(0) = x_3(1)$, ..., $x_{k-1}(0) = \vartheta_0(1)$.

Now, consider the following hypothesis:

H_0.   System (2.1), under the restriction $u(t) = v(t)$ has a unique solution

$$x(t_0 + \theta) = \varphi(\theta, \varphi_0, v), \qquad \theta \in [0, T] \tag{2.14}$$

in the interval $[0, T]$.

Under H_0 the following result was proven in Garcia-Ramirez et al. (2016a), using as basic argument that the solution of part of the equations can be recovered as copies, delayed, of the solutions of (2.1).

Theorem 2.1 *Consider system (2.1), with initial conditions $\varphi_0(\theta)$, and subject to the restriction $u(t) = v(t)$, $t \geq 0$, and assume that hypothesis H_0 is satisfied. Then, the solution (2.14) is obtained from the solution of system (2.5) in the interval $[t_0 + s, t_0 + T]$ if $T > k + s$,*

$$z_i(t_0 + s) = \varphi(s - i, \varphi_0, v), \ i = 0 \ldots, k + s,$$

and

$$\phi_i(t) = \dot{\varphi}(t - t_0 + i), \ i = 0, \ldots, s - 1.$$

As a corollary, one can consider the special case of a single delay as done in Example 2.2, thus getting the following result.

Corollary 2.1 *Given (2.9), consider the associated extended system (2.13). There always exists a proper initialization of (2.13) so that the trajectories of both systems coincide on the time interval $(k - 1) \leq t < k$.*

This means that it is possible to group the solution of (2.9) on the interval $(k - 1) \leq t < k$.

2.2   Non-independence of the Inputs of the Extended System

It has to be noticed that $u(t - k)$ and $x(t - k)$ are not independent of $u(t)$ and $x(t)$, so that attention must be paid when referring to the representation (2.3). An example is given hereafter.

Example 2.3 Consider, for instance, the two systems Σ_1 and Σ_2

$$\Sigma_1 : \begin{cases} \dot{x}_{1,[0]} = (x_{2,[0]}^2 + 1)u_{[0]} \\ \dot{x}_{2,[0]} = u_{[0]}(-1) \end{cases} , \qquad \Sigma_2 : \begin{cases} \dot{x}_{1,[0]} = (x_{2,[0]}^2 + 1)u_1 \\ \dot{x}_{2,[0]} = u_2. \end{cases}$$

It is easily seen that they represent the same dynamics whenever $u_1 = u_{[0]}$ and $u_2 = u_{[0]}(-1)$. So, Σ_2 has less constraints and better properties than Σ_1 regarding, for instance, feedback linearization: as a matter of fact while Σ_2 can be linearized via a regular static state feedback by setting

$$\begin{aligned} u_1 &= \frac{1}{x_{2,[0]}^2 + 1}v_1 \\ u_2 &= v_2, \end{aligned} \qquad (2.15)$$

there is no regular static state feedback which fully linearizes Σ_1, since $u_{[0]}(-1)$ is no more independent of $u_{[0]}$. In fact, setting

$$u_{[0]} = \frac{1}{x_{2,[0]}^2 + 1}v_{[0]}$$

would immediately imply that

$$u_{[0]}(-1) = \frac{1}{x_{2,[0]}^2(-1) + 1}v_{[0]}(-1).$$

Still, the feedback (2.15) can be implemented on the time-delay system Σ_1 on any time interval $[2kT, (2k + 1)T)$ and can switch on $u(t) = v(t)$ for any $t \in [(2k + 1)T, (2k + 2)T)$, with the initialization $u_{[0]}(-1) = 0$. This switching scheme ensures a linear behavior for the given dynamics on any interval $[2kT, (2k + 1)T], k \geq 0$.

The conclusion at this stage is that one cannot neglect the links between the control/state variables and their delayed signals. As shown hereafter, this is one of the problems that can be overcome by considering the differential representation of the given time-delay system. Such a representation naturally takes into account through the shift operator the link of a given variable with its delayed terms.

2.3 The Differential Form Representation

One of the peculiarities of nonlinear time-delay systems is the fact that when analyzing their dynamics, one has to refer to two different kind of operations with respect to time: time differentiation and time shift.

The simultaneous action of shift and differentiation determines difficulties which are peculiar to time-delay systems. A simple case is illustrated through the following example.

Example 2.4 Consider the two following nonlinear time-delay systems:

$$\Sigma_1 : \begin{cases} \dot{x}_{1,[0]} = x_{2,[0]}^3(-1) + x_{2,[0]} + x_{2,[0]}^3 \\ \qquad\quad +x_{2,[0]}(-1) \\ \dot{x}_{2,[0]} = u_{1,[0]} \\ y_{1,[0]} = x_{1,[0]} \end{cases} \qquad \Sigma_2 : \begin{cases} \dot{x}_{1,[0]} = x_{2,[0]}(-1)x_{2,[0]} \\ \dot{x}_{2,[0]} = u_{2,[0]} \\ y_{2,[0]} = x_{1,[0]}. \end{cases}$$

The input–output behavior is obtained by considering, in these two cases, the second-order derivatives of the output maps. We easily get that

$$\ddot{y}_{1,[0]} = \left(3x_{2,[0]}^2(-1) + 1\right) u_{1,[0]}(-1) + \left(3x_{2,[0]}^2 + 1\right) u_{1,[0]}$$
$$\ddot{y}_{2,[0]} = x_{2,[0]}(-1)u_{2,[0]} + u_{2,[0]}(-1)x_{2,[0]}. \tag{2.16}$$

While in the first case the feedback $u_{1,[0]} = \frac{1}{3x_{2,[0]}^2+1}v_{1,[0]}$ linearizes the input–output behavior, that is, $\ddot{y}_{1,[0]} = v_{1,[0]} + v_{1,[0]}(-1)$, there is instead no regular static state feedback which allows to solve the same problem for the second system.

The difference between these two cases can be understood through the use of the differential form representation, where the shifts are represented by the δ operator, and the representation becomes linear.

More precisely, consider the time-delay system (2.1), (2.2), and recall that, using the notation introduced in Sect. 1.5, for any $k \geq 0$, $dx(t-k) = dx_{[0]}(-k) = \delta^k dx_{[0]}$ and similarly, for any $\ell \geq 0$, $du(t-\ell) = du_{[0]}(-\ell) = \delta^\ell du_{[0]}$. Through standard computations one gets that such a differential form representation is given by

$$d\dot{x}_{[0]} = f(x_{[s]}, u_{[s]}, \delta)dx_{[0]} + \sum_{j=1}^{m} g_{1,j}(x_{[s]}, \delta)du_{[0],j}$$
$$dy_{[0]} = h(x_{[s]}, \delta)dx_{[0]}, \tag{2.17}$$

where $f(x_{[s]}, u_{[s]}, \delta)$ is a $n \times n$ matrix representing the differential with respect to the state variable and is given by

$$f(x_{[s]}, u_{[s]}, \delta) = \sum_{\ell=0}^{s} \frac{\partial F(x_{[s]})}{\partial x_{[0]}(-\ell)} \delta^\ell + \sum_{j=1}^{m} \sum_{\ell=0}^{s} \sum_{i=0}^{l} u_{[0]}(-i) \frac{\partial G_{ji}(x_{[s]})}{\partial x_{[0]}(-\ell)} \delta^\ell,$$

$g_{1,j}(\mathbf{x}_{[s]}, \delta) = \sum_{\ell=0}^{l} g_{1,j}^{\ell}(\mathbf{x})\delta^{\ell}$ is a $n \times 1$ vector representing the differential of the dynamics with respect to the control u_j, and given by

$$g_{1,j}(\mathbf{x}_{[s]}, \delta) = \sum_{i=0}^{l} G_{j,i}(\mathbf{x}_{[s]})\delta^i, \qquad j \in [1, m].$$

Finally, $h_j(\mathbf{x}_{[s]}, \delta) = \sum_{\ell=0}^{s} h_j^{\ell}(\mathbf{x})\delta^{\ell}$ is an $1 \times n$ vector representing the differential of the output, and is given by

$$h_j(\mathbf{x}_{[s]}, \delta) = \sum_{i=0}^{s} \frac{\partial H_j(\mathbf{x}_{[s]})}{\partial \mathbf{x}_{[0]}(-i)}\delta^i, \qquad j \in [1, p].$$

Consider again Example 2.4. The differential form representation of Σ_1 is

$$\begin{cases} d\dot{x}_{1,[0]} = (3x_{2,[0]}^2 + 1 + 3x_{2,[0]}^2(-1)\delta + \delta)dx_{2,[0]} \\ d\dot{x}_{2,[0]} = du_{1,[0]} \\ dy_{1,[0]} = dx_{1,[0]} \end{cases}$$

and the differentials of the derivatives of the output are

$$d\dot{y}_{1,[0]} = d\dot{x}_{1,[0]} = (3x_{2,[0]}^2 + 1 + 3x_{2,[0]}^2(-1)\delta + \delta)dx_{2,[0]}$$
$$d\ddot{y}_{1,[0]} = (6x_{2,[0]}\dot{x}_{2,[0]} + 6x_{2,[0]}(-1)\dot{x}_{2,[0]}(-1)\delta)dx_{2,[0]}$$
$$+ (3x_{2,[0]}^2 + 1 + 3x_{2,[0]}^2(-1)\delta + \delta)d\dot{x}_{2,[0]}.$$

With some technical manipulations, using the fact that $f(-1)\delta = \delta f(0)$, one then gets that

$$d\ddot{y}_{1,[0]} = (1 + \delta)\left(6u_{1,[0]}x_{2,[0]}dx_{2,[0]} + (3x_{2,[0]}^2 + 1)du_{1,[0]}\right).$$

Since the left side is an exact differential, also the right-hand side is an exact differential and it is possible to find the solution of

$$6u_{1,[0]}x_{2,[0]}dx_{2,[0]} + (3x_{2,[0]}^2 + 1)du_{1,[0]} = d\left[(3x_{2,[0]}^2 + 1)u_{1,[0]}\right] = dv_{1,[0]},$$

that is,

$$u_{1,[0]} = \frac{1}{3x_{2,[0]}^2 + 1}v_{1,[0]}.$$

With such a feedback $dy_{1,[0]} = dv_{1,[0]} + dv_{1,[0]}(-1)$ as expected.

Consider now system Σ_2. Its differential representation is

$$\begin{cases} d\dot{x}_{1,[0]} = \left(x_{2,[0]}(-1) + x_{2,[0]}\delta\right) dx_{2,[0]} \\ d\dot{x}_{2,[0]} = du_{2,[0]} \\ dy_{2,[0]} = dx_{1,[0]} \end{cases}$$

and the differentials of the output derivatives are

$$\begin{aligned} d\dot{y}_{2,[0]} &= d\dot{x}_{1,[0]} = \left(x_{2,[0]}(-1) + x_{2,[0]}\delta\right) dx_{2,[0]} \\ d\ddot{y}_{2,[0]} &= \left(\dot{x}_{2,[0]}(-1) + \dot{x}_{2,[0]}\delta\right) dx_{2,[0]} + \left(x_{2,[0]}(-1) + x_{2,[0]}\delta\right) d\dot{x}_{2,[0]} \\ &= \left(x_{2,[0]}(-1) + x_{2,[0]}\delta\right) du_{2,[0]} + \left(u_{2,[0]}(-1) + u_{2,[0]}\delta\right) dx_{2,[0]}. \end{aligned}$$

Since the coefficient of $du_{2,[0]}$ cannot be factorized in $c_0(\delta)c_1(\mathbf{x})$, there is no static state feedback which can achieve input–output linearization.

2.4 Generalized Lie Derivative and Generalized Lie Bracket

When dealing with nonlinear systems, Lie derivatives and Lie Brackets are standard tools used in many contexts (Isidori 1995). As well known the Lie derivative represents the derivative of a function along a given trajectory. When moving to the time-delay context, however, several aspects should be taken into account.

As a first comment, consider again the dynamics (2.1) and consider $g_{1,j}(\mathbf{x}, \delta)$ in (2.17) which is thus associated to the differential representation of (2.1).

Accordingly, it is immediate to understand that $\delta^k g_{1,j}(\mathbf{x}, \delta)$ will be associated to the differential representation of

$$\dot{\mathbf{x}}_{[0]}(-k) = F(\mathbf{x}_{[s]}(-k)) + \sum_{i=0}^{s}\sum_{j=1}^{m} G_{ji}(\mathbf{x}_{[s]}(-k))u_{j,[0]}(-i - k). \quad \bullet \quad (2.18)$$

Such a reasoning can be generalized to any element $r(\mathbf{x}, \delta) = \sum_{\ell=0}^{s} r^{\ell}(\mathbf{x})\delta^{\ell}$, so that if $r(\mathbf{x}, \delta)$ is associated to the differential representation of (2.1), then $\delta^k r(\mathbf{x}, \delta) = \sum_{\ell=0}^{s} r^{\ell}(\mathbf{x}_{[0]}(-k))\delta^{\ell+k}$ will be associated to the differential representation of (2.18).

One thus gets the following infinite dimensional matrix:

$$\begin{matrix} \frac{\partial}{\partial \mathbf{x}_{[0]}} \rightarrow \\ \frac{\partial}{\partial \mathbf{x}_{[0]}(-1)} \rightarrow \\ \vdots \\ \frac{\partial}{\partial \mathbf{x}_{[0]}(-s)} \rightarrow \\ \vdots \end{matrix} \left\{ \begin{matrix} r^0 & r^1 & \cdots & r^s & 0 & \cdots \\ 0 & r^0(-1) & \cdots & r^{s-1}(-1) & r^s(-1) & 0 & \ddots \\ \vdots & \ddots & \ddots & \vdots & \vdots & \ddots & \cdots \\ 0 & \cdots & 0 & r^0(-s) & r^1(-s) & \cdots & \ddots \\ \vdots & \vdots & \vdots & \vdots & \vdots & \vdots & \cdots \end{matrix} \right\}. \qquad (2.19)$$

Despite the infinite dimensionality of the state-space $x_e = (x^T(t), x^T(t-1), x^T(t-2), \ldots)^T$ the columns are generated by a finite number of elements, depending on a finite number of variables. This fact will turn to play a crucial role in the definitions of Lie derivative and Lie Bracket for time-delay systems introduced in Califano et al. (2011a), Califano and Moog (2017), and which are given hereafter.

Definition 2.1 (*Generalized Lie derivative*) Given the function $\tau(\mathbf{x}_{[p,s]})$ and the submodule element $r(\mathbf{x}, \delta) = \sum_{j=0}^{\bar{s}} r^j(\mathbf{x})\delta^j \in \mathcal{K}^{*n}(\delta]$, the Generalized Lie derivative $L_{r^\mu(\mathbf{x})}\tau(\mathbf{x}_{[p,s]})$, $\mu \in [0, \bar{s}]$ is defined as

$$L_{r^\mu(\mathbf{x})}\tau(\mathbf{x}_{[p,s]}) = \sum_{l=-p}^{\mu} \frac{\partial \tau(\mathbf{x}_{[p,s]})}{\partial \mathbf{x}_{[0]}(-l)} r^{\mu-l}(\mathbf{x}(-l)). \qquad (2.20)$$

Remark 2.1 In a delay-free context, one would have that $p = s = \bar{s} = 0$ and the Generalized Lie derivative would reduce to

$$L_{r^0(x)}\tau(x) = \frac{\partial \tau(x)}{\partial x} r^0(x),$$

which is exactly the standard Lie derivative of τ along r^0.

Definition 2.2 (*Generalized Lie Bracket*) Let $r_q(\mathbf{x}, \delta) = \sum_{j=0}^{\bar{s}} r_q^j(\mathbf{x})\delta^j \in \mathcal{K}^{*n}(\delta]$, $q = 1, 2$. For any $k, l \geq 0$, the Generalized Lie Bracket $[r_1^k(\cdot), r_2^l(\cdot)]_{E_i}$, on $IR^{(i+1)n}$, $i \geq 0$, is defined as

$$[r_1^k(\cdot), r_2^l(\cdot)]_{E_i} = \sum_{j=0}^{i} \left([r_1^{k-j}, r_2^{l-j}]_E\right)_{(\mathbf{x}(-j))}^T \frac{\partial}{\partial \mathbf{x}_{[0]}(-j)}, \qquad (2.21)$$

where

$$[r_1^k(\cdot), r_2^l(\cdot)]_E = \left(L_{r_1^k(\mathbf{x})} r_2^l(\mathbf{x}) - L_{r_2^l(\mathbf{x})} r_1^k(\mathbf{x})\right). \qquad (2.22)$$

Remark 2.2 The Generalized Lie derivative as defined by (2.20) is the Lie derivative of $\tau(\mathbf{x}_{[p,s]})$ along

$$(r^{\mu+p}(+p), \ldots, r^{\mu}(0), r^{\mu-1}(-1), \ldots, r^0(-\mu), 0)^T.$$

The latter is embedded in

$$\Delta_{[p,q]} = \mathrm{span}_{\mathcal{K}^*} \begin{pmatrix} \mathbf{r}^0(\mathbf{x}(p)) & \cdots & \mathbf{r}^\ell(\mathbf{x}(p)) & 0 & 0 \\ 0 & \ddots & & \ddots & 0 \\ 0 & & 0 & \mathbf{r}^0(\mathbf{x}(-q)) & \cdots & \mathbf{r}^\ell(\mathbf{x}(-q)) \end{pmatrix},$$

where $\mathbf{r}^i(\mathbf{x}) = (r_1^i, \ldots, r_j^i)$ and $q > \mu$. Accordingly, assuming without loss of generality $k \geq l$, the Generalized Lie Bracket $[r_1^k(\cdot), r_2^l(\cdot)]_{E_i}$ is defined starting from the standard Lie Bracket

$$\left[\begin{pmatrix} 0 \\ r_1^s(s-k) \\ \vdots \\ r_1^k(0) \\ \vdots \\ r_1^0(-k) \\ 0 \end{pmatrix} \begin{pmatrix} r_2^s(s-l) \\ \vdots \\ r_2^l(0) \\ \vdots \\ r_2^0(-l) \\ 0 \\ 0 \end{pmatrix} \right] = \begin{pmatrix} \tau^{k+s-l}(s-l) \\ \vdots \\ \vdots \\ \tau^k(0) \\ \vdots \\ \tau^0(-k) \\ 0 \end{pmatrix}.$$

In fact, $[r_1^k(\cdot), r_2^l(\cdot)]_{E_i} = \sum\limits_{j=0}^{min(k,i)} (\tau^{k-j}(-j))^T \frac{\partial}{\partial \mathbf{x}(-j)}.$

The Generalized Lie Brackets (2.21) are associated to $\Delta_{[p,q]}$ defined above. In the special case of causal submodules (which lead to consider $\Delta_{[0,q]}$), they have shown to characterize the 0-integrability conditions, that is, when the $\Delta^\perp(\delta)$ of rank $n-j$ is generated by $n-j$ exact and independent differentials $d\lambda_\mu(\mathbf{x}) = \Lambda_\mu(\mathbf{x}, \delta)d\mathbf{x}_{[0]}$, $\mu \in [1, n-j]$ (Califano et al. 2011a). In order to give the conditions on integrability directly on the submodule

$$\Delta(\delta) = \mathrm{span}_{\mathcal{K}^*(\delta)}\{r_1(\mathbf{x}, \delta), \ldots, r_j(\mathbf{x}, \delta)\},$$

we need to refer to the definition of Polynomial Lie Bracket and accordingly to a more general definition of Lie Bracket.

Definition 2.3 (*Lie Bracket*) Given $r_i(\mathbf{x}_{[s_i, s]}, \delta) \in \mathcal{K}^{*n}(\delta), i = 1, 2$, the Lie Bracket

$$[r_1(\mathbf{x}_{[s_1, s]}, \delta), r_2(\mathbf{x}_{[s_2, s]}, \delta)],$$

is a $(4s + s_1 + s_2 + 1)$-uple of polynomial vectors $r_{12,j}(\mathbf{x}, \delta)$, defined as

$$r_{12,j}(\mathbf{x}, \delta) = \sum_{\ell=-s_1}^{2s+s_1} [r_1^{\ell+s_1-j}, r_2^{\ell}]_{E_0} \delta^{\ell+s_1}, \quad j \in [-2s, 2s + s_1 + s_2]. \qquad (2.23)$$

Recalling that a polynomial vector $r_1(\mathbf{x}_{[s_i,s]}, \delta)$ acts on a function $\epsilon(t)$ and denoting its image as $\mathbf{R}_1(\mathbf{x}_{[s_1,s]}, \epsilon) := \sum_{j=0}^{s} r_1^j(\mathbf{x})\epsilon(-j)$, the Polynomial Lie Bracket is then defined as follows:

Definition 2.4 (*Polynomial Lie Bracket*) Given $r_i(\mathbf{x}_{[s_i,s]}, \delta) \in \mathcal{K}^{*n}(\delta)$, $i = 1, 2$, the Polynomial Lie Bracket $[\mathbf{R}_1(\mathbf{x}, \epsilon), r_2(\mathbf{x}, \delta)]$ is defined as

$$[\mathbf{R}_1(\mathbf{x}, \epsilon), r_2(\mathbf{x}, \delta)] := ad_{\mathbf{R}_1(\mathbf{x}_{[s_1,s]}, \epsilon)} r_2(\mathbf{x}_{[s_2,s]}, \delta) =$$

$$\dot{r}_2(\mathbf{x}, \delta)|_{\dot{x}_{[0]} = \mathbf{R}_1(\mathbf{x}, \epsilon)} \delta^{s_1} - \sum_{k=0}^{s_1+s} \frac{\partial \mathbf{R}_1(\mathbf{x}_{[s_1,s]}, \epsilon)}{\partial \mathbf{x}(s_1 - k)} \delta^k r_2(\mathbf{x}(s_1), \delta).$$

With some abuse, the Polynomial Lie Bracket and the standard Lie Bracket are both denoted by $[.,.]$. No confusion is possible, since in the Polynomial Lie Bracket, some $\epsilon(i)$ will always be present inside the brackets.

Some comments

As noted in Califano and Moog (2017), the link between the Lie Bracket (2.23) and the Generalized Lie Bracket (2.21) can be easily established by noting that $r_{12,j}(\mathbf{x}, \delta)$ in (2.23) is given by

$$r_{12,j}(\mathbf{x}, \delta) = \mathbf{I}(\delta) \left([r_1^{2(s+s_1)-j}, r_2^{2s+s_1}]_{E_{2s+s_1}} |_{\mathbf{x}(2(s+s_1))} \right),$$

where

$$\mathbf{I}(\delta) = \left(I_n \delta^{2(s+s_1)}, \cdots, I_n \delta, I_n \right).$$

Furthermore, the $r_{12,j}(\mathbf{x}, \delta)$'s also characterize the Polynomial Lie Bracket since one easily gets that

$$[\mathbf{R}_1(\mathbf{x}, \epsilon), r_2(\mathbf{x}, \delta)] = \sum_{j=-2s}^{2s+s_1+s_2} r_{12,j}(\mathbf{x}, \delta)\epsilon(j). \qquad (2.24)$$

Finally, in the delay-free case, the Polynomial Lie Bracket reduces (up to $\epsilon(0)$) to the standard Lie Bracket. In fact

$$[\mathbf{R}_1(\mathbf{x}, \epsilon), r_2(\mathbf{x}, \delta)] = [r_1^0(x)\epsilon(0), r_2^0(x)] = [r_1^0, r_2^0]\epsilon(0).$$

Instead, if delays are present, $[\mathbf{R}_1(\mathbf{x}, \epsilon), r_2(\mathbf{x}, \delta)]$ immediately enlightens some important differences with respect to the delay-free case, such as the loss of validity of the Straightening theorem. In fact, since the term depending on δ undergoes a different kind of operation with respect to the term depending on ϵ, starting from $r(\mathbf{x}, \delta)$ and its corresponding image $\mathbf{R}(\mathbf{x}, \epsilon)$, in general,

$$\dot{r}(\mathbf{x}, \delta)|_{\dot{\mathbf{x}}_{[0]}=\mathbf{R}(\mathbf{x}, \epsilon)} \delta^{s_1} \neq \sum_{k=0}^{s_1+s} \frac{\partial \mathbf{R}(\mathbf{x}_{[s_1,s]}, \epsilon)}{\partial \mathbf{x}(s_1 - k)} \delta^k r(\mathbf{x}(s_1), \delta),$$

which yields that, in general, $[r(\mathbf{x}, \delta), r(\mathbf{x}, \delta)] \neq 0$. For instance, consider $r(\mathbf{x}, \delta) = \begin{pmatrix} x_2(-1) \\ 1 \end{pmatrix}$. Then $\mathbf{R}(\mathbf{x}, \epsilon) = \begin{pmatrix} x_2(-1) \\ 1 \end{pmatrix} \epsilon(0)$ and

$$[\mathbf{R}(\mathbf{x}, \epsilon), r(\mathbf{x}, \delta)] = \begin{pmatrix} \epsilon(-1) - \epsilon(0)\delta \\ 0 \end{pmatrix} \neq 0.$$

Accordingly,

$$[r(\mathbf{x}, \delta), r(\mathbf{x}, \delta)] = \left\{ \begin{pmatrix} 1 \\ 0 \end{pmatrix}, \begin{pmatrix} -\delta \\ 0 \end{pmatrix} \right\}.$$

2.5 Some Remarks on the Polynomial Lie Bracket

Let us first examine some properties of the Polynomial Lie Bracket discussed in Califano and Moog (2017).

Property 2.1 (*Anticommutativity*) Assume without loss of generality, $s_2 \geq s_1$, then for any integer j,

$$\frac{\partial [\mathbf{R}_1(\mathbf{x}, \epsilon), r_2(\mathbf{x}, \delta)]}{\partial \epsilon(s_1 - j)} \delta^{s_2 - s_1 + j + |j|} = -\frac{\partial [\mathbf{R}_2(\mathbf{x}, \epsilon), r_1(\mathbf{x}, \delta)]}{\partial \epsilon(s_2 + j)} \delta^{|j|} \qquad (2.25)$$

Property 2.2 Given for $i = 1, 2$, $\bar{r}_i(\mathbf{x}_{[\bar{s}_i, s]}, \delta) = r_i(\mathbf{x}_{[s_i, s]}, \delta)\beta_i(\mathbf{x}_{[s_i, s]}, \delta)$, then

$$[\bar{\mathbf{R}}_1(\mathbf{x}, \epsilon), \bar{r}_2(\mathbf{x}, \delta)]\delta^{s_1 - \bar{s}_1} =$$

$$[\mathbf{R}_1(\mathbf{x}, \bar{\epsilon}), r_2(\mathbf{x}, \delta)]_{\bar{\epsilon}=\beta_1(\mathbf{x}, \epsilon)}\hat{\beta}_2 + r_2(\mathbf{x}, \delta)\alpha_2 - r_1(\mathbf{x}, \delta)\alpha_1 \qquad (2.26)$$

with $\hat{\beta}_2 = \beta_2(\mathbf{x}(s_1), \delta)$, $\alpha_1 = \sum_{k=0}^{s+s_1} \frac{\partial \beta_1(\mathbf{x}, \epsilon)}{\partial \mathbf{x}(s_1 - k)} \delta^k \bar{r}_2(\mathbf{x}(s_1), \delta)$, and $\alpha_2 = \dot{\beta}_2(\mathbf{x}, \delta)|_{\dot{\mathbf{x}}=\bar{\mathbf{R}}_1(\mathbf{x}, \epsilon)} \delta^{s_1}$.

Property 2.3 The repeated Polynomial Lie Bracket obtained by setting $\epsilon = 1$ is given by

$$ad_{\mathbf{R}_1(\mathbf{x},1)}^k (r_2(\mathbf{x},\delta)\alpha) = \sum_{j=0}^{k} \binom{k}{j} ad_{\mathbf{R}_1(\mathbf{x},1)}^{k-j} (r_2(\mathbf{x},\delta))\, \alpha^{(j)}|_{\dot{\mathbf{x}}=\mathbf{R}_1(\mathbf{x},1)}. \qquad (2.27)$$

It is important to point out that for delay-free systems one recovers the standard properties of Lie Brackets. In fact, if $r_i(\mathbf{x},\delta) = r_i^0(x)$, for $i = 1, 2$, then $\mathbf{R}_i(\mathbf{x},\epsilon) = r_i^0(x)\epsilon(0)$ and

$$\frac{\partial[\mathbf{R}_1(\mathbf{x},\epsilon), r_2(\mathbf{x},\delta)]}{\partial\epsilon(0)} = [r_1^0, r_2^0] = -[r_2^0, r_1^0] = -\frac{\partial[\mathbf{R}_2(\mathbf{x},\epsilon), r_1(\mathbf{x},\delta)]}{\partial\epsilon(0)}$$

whereas letting $\bar{r}_i(\mathbf{x},\delta) = r_i^0(x)\beta_i(x)$, then $\bar{\mathbf{R}}_i(\mathbf{x},\epsilon) = r_i^0(x)\beta_i(x)\epsilon(0)$ and

$$[\bar{\mathbf{R}}_1(\mathbf{x},\epsilon), \bar{r}_2(\mathbf{x},\delta)] = [r_1^0(x)\beta_1(x)\epsilon(0), r_2^0(x)\beta_2(x)]$$
$$= \left([r_1^0, r_2^0]\beta_2\beta_1 + r_2^0\alpha_2 - r_1^0\alpha_1\right)\epsilon(0)$$

with $\alpha_1 = \beta_2(L_{r_2^0}\beta_1)$ and $\alpha_2 = \beta_1(L_{r_1^0}\beta_2)$.

Example 2.5 Consider for $i = 1, 2$, $r_i(\mathbf{x},\delta)$ given by

$$r_1(\mathbf{x},\delta) = \begin{pmatrix} x_1(1) \\ x_2\delta \end{pmatrix}, \quad r_2(\mathbf{x},\delta) = \begin{pmatrix} x_2\delta \\ x_1 \end{pmatrix}.$$

Then

$$\mathbf{R}_1(\mathbf{x},\epsilon) = \begin{pmatrix} x_1(1)\epsilon(0) \\ x_2\epsilon(-1) \end{pmatrix}, \quad \mathbf{R}_2(\mathbf{x},\epsilon) = \begin{pmatrix} x_2\epsilon(-1) \\ x_1\epsilon(0) \end{pmatrix}.$$

Accordingly, since $s_1 = 1$, $s_2 = s = 0$,

$$[\mathbf{R}_1(\mathbf{x},\epsilon), r_2(\mathbf{x},\delta)] = \begin{pmatrix} x_2\epsilon(-1)\delta \\ x_1(1)\epsilon(0) \end{pmatrix}\delta - \begin{pmatrix} \epsilon(0)x_2(1)\delta \\ \epsilon(-1)x_1\delta \end{pmatrix}$$
$$= -\begin{pmatrix} 0 \\ x_1\delta \end{pmatrix}\epsilon(0) + \begin{pmatrix} x_2\delta^2 - x_2(1)\delta \\ x_1(1)\delta \end{pmatrix}\epsilon(1)$$
$$= r_{12,0}(\mathbf{x},\delta)\epsilon(0) + r_{12,1}(\mathbf{x},\delta)\epsilon(1).$$

One can easily verify that

$$r_{12,0}(\mathbf{x},\delta) = -\begin{pmatrix} 0 \\ x_1 \end{pmatrix}\delta = \sum_{\ell=-1}^{1} [r_1^{\ell+1}, r_2^{\ell}]_{E_0}\delta^{\ell+1}$$

$$r_{12,1}(\mathbf{x},\delta) = \begin{pmatrix} -x_2(1) \\ x_1(1) \end{pmatrix}\delta + \begin{pmatrix} x_2 \\ 0 \end{pmatrix}\delta^2 = \sum_{\ell=-1}^{1} [r_1^{\ell}, r_2^{\ell}]_{E_0}\delta^{\ell+1},$$

which confirms Eq. (2.23).

Analogously, $[\mathbf{R}_2(\mathbf{x}, \epsilon), r_1(\mathbf{x}, \delta)] = \begin{pmatrix} x_2(1) - x_2\delta \\ -x_1(1) \end{pmatrix} \epsilon(0) + \begin{pmatrix} 0 \\ x_1\delta \end{pmatrix} \epsilon(1)$ and it is again easily verified that (2.25) holds true (with the indices exchanged since $s_1 > s_2$). In fact,

$$\frac{\partial[\mathbf{R}_2(\mathbf{x}, \epsilon), r_1(\mathbf{x}, \delta)]}{\partial \epsilon(0)} \delta = \begin{pmatrix} x_2(1) - x_2\delta \\ -x_1(1) \end{pmatrix} \delta = -\frac{\partial[\mathbf{R}_1(\mathbf{x}, \epsilon), r_2(\mathbf{x}, \delta)]}{\partial \epsilon(1)}$$

$$\frac{\partial[\mathbf{R}_2(\mathbf{x}, \epsilon), r_1(\mathbf{x}, \delta)]}{\partial \epsilon(1)} \delta = \begin{pmatrix} 0 \\ x_1\delta \end{pmatrix} \delta = -\frac{\partial[\mathbf{R}_1(\mathbf{x}, \epsilon), r_2(\mathbf{x}, \delta)]}{\partial \epsilon(0)} \delta.$$

Since the derivative of a differential is equal to the differential of the derivative one can easily compute the differential of the kth derivative.

As already underlined, the Polynomial Lie Bracket is intrinsically linked to the standard Lie Bracket when we consider delay-free systems. Since to the standard Lie Bracket it has been given a precise geometric interpretation, one may wonder if something could be said also in the delay case. Let us, in fact, recall that in the delay-free case the geometric interpretation of the Lie Bracket can be easily obtained by considering a simple example given in Spivak (1999) of a two-input driftless system of the form

$$\dot{x}(t) = g_1(x(t))u_1(t) + g_2(x(t))u_2(t).$$

If the system were linear, that is, $g_1(x)$ and $g_2(x)$ were constant vectors, the application of an input sequence of the form $[(0, 1), (1, 0), (0, -1), (-1, 0)]$ where each control acts exactly for a time h, would bring the state back to the starting point. In the nonlinear case instead it was shown that such a sequence brings the system into a final point x_f different from the starting one x_0 and that the Lie Bracket $[g_2, g_1]$ exactly identifies the direction which should be taken to go back to x_0 from x_f. In fact, if one carries out the computation it turns out that the first-order derivative of the flow in the origin is zero, while its second-order derivative evaluated again in the origin is exactly twice the bracket $[g_2, g_1]$. As a by-product in the special case of a single-input driftless and delay-free system then using a constant control allows to move forward or backward on a unique integral manifold of the considered control vector field, and this can be easily proven by considering that $[g, g] = 0$.

In the time-delay context, already for a single-input system the Polynomial Lie Bracket is not, in general, identically zero. Using analogous arguments as in the delay-free case, as already discussed in Sect. 1.3, it follows that already in the single-input case, in fact, it is not, in general, true that one moves forward and backward on a unique integral manifold when delays are present. To formally show this, let us go back to the single-input time-delay system (1.8)

$$\dot{x}(t) = g(x(t), x(t - \tau))u(t) = \begin{pmatrix} x_2(t - \tau) \\ 1 \end{pmatrix} u(t) \tag{2.28}$$

and consider the dynamics over four steps of magnitude τ when the control sequence $[1, 0, -1, 0]$ is applied and the switches occur every τ. Then one gets that over the four steps the dynamics reads

$$\begin{aligned}
\dot{x}(t) &= g(x(t), x(t - \tau))u(t) \\
\dot{x}(t - \tau) &= g(x(t - \tau), x(t - 2\tau))u(t - \tau) \\
\dot{x}(t - 2\tau) &= g(x(t - 2\tau), x(t - 3\tau))u(t - 2\tau) \\
\dot{x}(t - 3\tau) &= g(x(t - 3\tau), x(t - 4\tau))u(t - 3\tau)
\end{aligned} \tag{2.29}$$

and due to the input sequence, it can be rewritten in the form

$$\dot{z}(t) = g_1(z(t))u_1(t) + g_2(z(t))u_2(t), \tag{2.30}$$

where $z_1(t) = x(t)$, $z_2(t) = x(t - \tau)$, $z_3(t) = x(t - 2\tau)$, $z_4(t) = x(t - 3\tau)$, $u_1(t) = u(t - \tau) = -u(t - 3\tau)$, and $u_2(t) = u(t) = -u(t - 2\tau)$. In (2.30)

$$g_1(z) = \begin{pmatrix} 0 \\ g(z_2, z_3) \\ 0 \\ -g(z_4, c_0) \end{pmatrix}, \quad g_2(z) = \begin{pmatrix} g(z_1, z_2) \\ 0 \\ -g(z_3, z_4) \\ 0 \end{pmatrix},$$

with c_0 the initial condition of x on the interval $[-4\tau, -3\tau)$. Of course not all the trajectories of $z_1(t)$ in (2.30) will be trajectories of $x(t)$ in (2.29), whereas all the trajectories of $x(t)$ for $t \in [0, 4\tau)$ in (2.29) can be recovered as trajectories of $z_1(t)$ in (2.30) for $t \in [0, 4\tau)$, whenever the system is initialized with constant initial conditions.

Mimicking the delay-free case, one should then apply the input sequence $[(0, 1), (1, 0), (0, -1), (-1, 0)]$ to the system. This can be achieved by considering $u(t) = 1$ for $t \in [0, \tau)$, $u(t) = 0$ for $t \in [\tau, 2\tau)$, $u(t) = -1$ for $t \in [2\tau, 3\tau)$, and $u(t) = 0$ for $t \in [3\tau, 4\tau)$, with the initialization $u(t) = 0$ for $t \in [-\tau, 0)$. Such an example shows immediately that the second-order derivative in 0 is characterized by

$$\begin{aligned}
[g_1, g_2] &= \left[\begin{pmatrix} 0 \\ g(z_2, z_3) \\ 0 \\ -g(z_4, c_0) \end{pmatrix}, \begin{pmatrix} g(z_1, z_2) \\ 0 \\ -g(z_3, z_4) \\ 0 \end{pmatrix} \right] \\
&= \left[\begin{pmatrix} 0 \\ g(x(t - \tau), x(t - 2\tau)) \\ 0 \\ -g(x(t - 3\tau), x(t - 4\tau)) \end{pmatrix}, \begin{pmatrix} g(x(t), x(t - \tau)) \\ 0 \\ -g(x(t - 2\tau), x(t - 3\tau)) \\ 0 \end{pmatrix} \right].
\end{aligned}$$

It is straightforward to note that the $\frac{\partial}{\partial x(t)}$ component of the Lie Bracket is given by $\frac{\partial g(x(t), x(t-\tau))}{\partial x(t-\tau)} g(x(t - \tau), x(t - 2\tau))$ which, in general, is nonzero, thus showing that the presence of a delay may generate a new direction that can be taken.

Such a result can be easily recovered by using the Polynomial Lie Bracket. In fact, one has that starting from $g_1(\mathbf{x}, \delta) = g(x(t), x(t - \tau))$, one considers $\mathbf{G}_1(\mathbf{x}, \epsilon) = g(x(t), x(t - \tau))\epsilon(0)$. Accordingly, the associated Polynomial Lie Bracket is

$$[\mathbf{G}_1(x, \epsilon), g_1(x, \delta)] = \dot{g}|_{\dot{x}_{[0]}=g(\mathbf{x}(0), \mathbf{x}(-1))\epsilon(0)} - \epsilon(0) \sum_{j=0}^{1} \frac{\partial g(\mathbf{x}(0), \mathbf{x}(-1))}{\partial \mathbf{x}(-j)} g(-j)$$

$$= \frac{\partial g(\mathbf{x}(0), \mathbf{x}(-1))}{\partial \mathbf{x}(-1)} g(\mathbf{x}(-1), \mathbf{x}(-2))(\epsilon(-1) - \epsilon(0)\delta)$$

$$= \begin{pmatrix} 1 \\ 0 \end{pmatrix} (\epsilon(-1) - \epsilon(0)\delta)$$

which is different from zero. Figure 1.2 shows the trajectories of the single-input system (2.28) controlled with a piecewise input which varies from 1 to -1 every 10 s, highlighting the difference with the delay-free case as discussed. It is also clear that in this framework the delay can be used as an additional control variable. This is left as an exercise to the reader, which can investigate on the effect of different delays on the system with the same output, as well as a system with a fixed delay and where the input changes its period.

2.6 The Action of Changes of Coordinates

Changes of coordinates play a fundamental role in the study of the structural properties of a given system. In the delay-free case, a classical example of their use is displayed by the decomposition in observable/reachable subsystems, following Kalman intuition. When dealing with time-delay systems, several problems arise when considering changes of coordinates.

Example 2.6 Consider, for instance, the system

$$\dot{x}(t) = x(t - 1) + u(t).$$

The map $z(t) = x(t) + x(t - 1)$ does not define a change of coordinates, since we are not able to express $x(t)$ as a function of $z(t)$ and a finite number of its delays. Nevertheless we can compute

$$\dot{z}(t) = z(t - 1) + u(t) + u(t - 1).$$

Example 2.7 Consider, for instance, the system

$$\dot{x}_1(t) = x_1(t - 1)x_2(t)$$
$$\dot{x}_2(t) = x_1(t).$$

The map $z_1(t) = x_1(t-1)$, $z_2(t) = x_2(t)$ leads to the system

$$\dot{z}_1(t) = z_1(t-1)z_2(t-1)$$
$$\dot{z}_2(t) = z_1(t+1),$$

where the causality property of the system has not been preserved.

The previous examples show that changes of coordinates should be defined with carefulness. To this end, we will consider bicausal changes of coordinates as defined in Márquez-Martínez et al. (2002), that is, changes of coordinates which are causal and admit a causal inverse map. Let us thus consider the mapping

$$\mathbf{z}_{[0]} = \varphi(\mathbf{x}_{[\alpha]}), \tag{2.31}$$

where $\alpha \in I\!N$ and $\varphi \in \mathcal{K}^n$.

Definition 2.5 Consider a system Σ in the state coordinates x. The mapping (2.31) is a local bicausal change of coordinates for Σ if there exists an integer $\ell \in I\!N$ and a function $\psi(\mathbf{z}_{[\ell]}) \in \mathcal{K}^n$ such that, assuming $\mathbf{z}_{[0]}$ and $\mathbf{x}_{[0]}$ defined for $t \geq -(\alpha + \ell)$, then $\psi(\varphi(\mathbf{x}_{[\alpha]}), \ldots, \varphi(\mathbf{x}_{[\alpha]}(-\ell))) = \mathbf{x}_{[0]}$ for $t \geq 0$.

Furthermore, if we consider the differential form representation of (2.31), which is given by

$$d\mathbf{z}_{[0]} = T(\mathbf{x}_{[\gamma]}, \delta)d\mathbf{x}_{[0]} = \sum_{j=0}^{s} T_j(\mathbf{x})\delta^j, \tag{2.32}$$

then the polynomial matrix $T(\mathbf{x}_{[\gamma]}, \delta) \in \mathcal{K}^{n \times n}(\delta)$ is unimodular and $\gamma \leq \alpha$, whereas its inverse $T^{-1}(\mathbf{z}, \delta)$ has polynomial degree $\ell \leq (n-1)\alpha$.

Under the change of coordinates (2.31), with differential representation (2.32), the differential representation (2.17) of the given system is transformed in the new coordinates into

$$d\dot{\mathbf{z}}_{[0]} = \tilde{f}(\mathbf{z}, \mathbf{u}, \delta)d\mathbf{z}_{[0]} + \sum_{j=1}^{m} \tilde{g}_{1,m}(\mathbf{z}, \delta)d\mathbf{u}_{[0]}$$
$$d\mathbf{y}_{[0]} = \tilde{h}(\mathbf{z}, \delta)d\mathbf{z}_{[0]} \tag{2.33}$$

with

$$\tilde{f}(\mathbf{z}, \mathbf{u}, \delta) = \left[\left(T(\mathbf{x}, \delta)f(\mathbf{x}, \mathbf{u}, \delta) + \dot{T}(\mathbf{x}, \delta)\right)T^{-1}(\mathbf{x}, \delta)\right]_{|\mathbf{x}_{[0]} = \phi^{-1}(\mathbf{z})}$$
$$\tilde{g}_{1,j}(\mathbf{z}, \delta) = \left(T(\mathbf{x}, \delta)g_{1,j}(\mathbf{x}, \delta)\right)_{|\mathbf{x}_{[0]} = \phi^{-1}(\mathbf{z})}, \tag{2.34}$$
$$\tilde{h}(\mathbf{z}, \delta) = \left(h(\mathbf{x}, \delta)T^{-1}(\mathbf{x}, \delta)\right)_{|\mathbf{x}_{[0]} = \phi^{-1}(\mathbf{z})}.$$

More generally, the effect of a bicausal change of coordinates on a submodule element is defined by the next result.

Proposition 2.1 *Under the bicausal change of coordinates* (2.31), *the causal sub-module element* $r(\mathbf{x}, \delta)$ *is transformed in* $\tilde{r}(\mathbf{z}, \delta)$ *given by*

$$\tilde{r}(\mathbf{z}, \delta) = [T(\mathbf{x}, \delta)r(\mathbf{x}, \delta)]_{\mathbf{x}_{[0]}=\phi^{-1}(\mathbf{z})} = \sum_{\ell=0}^{s+\bar{s}} \sum_{j=0}^{\ell} [T_j(\mathbf{x})r^{\ell-j}(\mathbf{x}(-j))]_{\mathbf{x}_{[0]}=\phi^{-1}(\mathbf{z})} \delta^{\ell}$$

with $s = deg(T(\mathbf{x}, \delta))$, $\bar{s} = deg(r(\mathbf{x}, \delta))$.

As a consequence, we are now able to characterize the effect of a bicausal change of coordinates on the Extended Lie Bracket and the Polynomial Lie Bracket. To this end, starting from the given change of coordinates and its differential representation (2.32), we have to consider the matrices

$$\Gamma^{l,i}(\mathbf{x}) = \begin{pmatrix} T^0(\mathbf{x}) \cdots & \cdots & \cdots & T^l(\mathbf{x}) \\ 0 & \ddots & & \vdots \\ 0 & 0 & T^0(\mathbf{x}(-i)) & \cdots & T^{l-i}(\mathbf{x}(-i)) \end{pmatrix},$$

where $T^j(\mathbf{x}) = 0$ for $j > s$. The following results hold true.

Lemma 2.1 *Let* $r_\beta(\mathbf{x}, \delta) = \sum_{j=0}^{\bar{s}} r_\beta^j(\mathbf{x})\delta^j$, $\beta = 1, 2$. *Under the bicausal change of coordinates* (2.31) *with differential representation* (2.32), *one has, for* $k \leq l$, $0 \leq i \leq l$

$$[\tilde{r}_1^k(\mathbf{z}), \tilde{r}_2^l(\mathbf{z})]_{E_i} = \left(\Gamma^{l,i}(\mathbf{x})[r_1^k(\mathbf{x}), r_2^l(\mathbf{x})]_{E_l}\right)_{|\mathbf{x}=\phi^{-1}(\mathbf{z})}.$$

Proposition 2.2 *Under the bicausal change of coordinates* (2.31), *the Polynomial Lie Bracket* $[R_1(\mathbf{x}, \epsilon), r_2(\mathbf{x}, \delta)]$ *defined starting from the causal submodule elements* $r_1(\mathbf{x}, \delta)$, $r_2(\mathbf{x}, \delta)$ *is transformed into*

$$[\tilde{R}_1(\mathbf{z}, \epsilon), \tilde{r}_2(\mathbf{z}, \delta)] = \sum_{j=-2\bar{s}}^{2\bar{s}} \tilde{r}_{12,j}(\mathbf{z}, \delta)\epsilon(j),$$

where

$$\tilde{r}_{12,j}(\mathbf{z}, \delta) = \sum_{\ell=0}^{2\bar{s}} [\tilde{r}_1^{\ell-j}, \tilde{r}_2^{\ell}]_{E_0} \delta^{\ell} = \sum_{\ell=0}^{2\bar{s}-j} [\tilde{r}_1^{\ell}, \tilde{r}_2^{\ell+j}]_{E_0} \delta^{j+\ell}$$

$$= \sum_{\ell=0}^{2\bar{s}-j} \left(\Gamma^{\ell+j,0}(\mathbf{x})[r_1^{\ell}(\mathbf{x}), r_2^{\ell+j}(\mathbf{x})]_{E_{\ell+j}}\right)_{|\mathbf{x}=\phi^{-1}(\mathbf{z})} \delta^{j+\ell}.$$

2.7   The Action of Static State Feedback Laws

Definition 2.6 Consider a system (2.1), an invertible instantaneous static state feedback is defined as

$$u(x(t), v(t)) = \alpha(x(t)) + \beta(x(t))v(t), \tag{2.35}$$

where $v(t)$ is a new input of dimension m and β is a square invertible matrix whose entries are meromorphic functions, so that (2.35) is invertible almost everywhere and one recovers $v(t)$ as a function of $u(t)$, that is,

$$v(x(t), u(t)) = [\beta(x(t))]^{-1}(-\alpha(x(t)) + u(t)).$$

In general, the class of instantaneous static state feedback laws is not rich enough to cope with the complexity of time-delay systems and to solve the respective control problems. Thus, delay-dependent state feedback laws are considered as well and have the same level of complexity as the system to be controlled.

Definition 2.7 Given the system (2.1), consider the feedback

$$u(\mathbf{x}, \mathbf{v}) = \alpha(x(t), \dots, x(t-\ell)) + \sum_{i=0}^{\ell} \beta_i(x(t), \dots, x(t-\ell))v(t-i), \tag{2.36}$$

which can be written in the compact form

$$u(\mathbf{x}, \mathbf{v}) = \alpha(\cdot) + \beta(x(t), \dots, x(t-\ell), \delta)v(t), \tag{2.37}$$

where v is a m-valued new input and

$$\beta(\mathbf{x}, \delta) = \sum_{i=0}^{i=\ell} \beta_i(x(t), \dots, x(t-\ell))\delta^i$$

is a δ-polynomial matrix.

The feedback (2.37) is said to be an invertible bicausal static state feedback if β is a unimodular polynomial matrix, i.e. it admits an inverse polynomial matrix β'.

It follows that v is a function of u as follows:

$$v(\mathbf{x}, \mathbf{u}) = [\beta'(\cdot, \delta)](-\alpha(\cdot) + u(t)). \tag{2.38}$$

Note that in the special case where $m = 1$, the invertibility of (2.36) necessarily yields

$$u(x(t)) = \alpha(x(t), \dots, x(t-\ell)) + \beta_0(x(t), \dots, x(t-\ell))v(t), \tag{2.39}$$

that is, the feedback law can depend only on $v(t)$. Only in the multi-input case, several time instants of $v(\cdot)$ may be involved. Referring to the differential representation, one gets that the differential of the feedback (2.36) is

$$
d\mathbf{u}_{[0]} = \sum_{j=0}^{\ell} \left(\frac{\partial \alpha(\mathbf{x})}{\partial \mathbf{x}(-j)} + \sum_{i=0}^{i=\ell} \frac{\partial \beta_i(\mathbf{x})}{\partial \mathbf{x}(-j)} \mathbf{v}(-i) \right) \delta^j d\mathbf{x}_{[0]} + \sum_{i=0}^{i=\ell} \beta_i(\mathbf{x}) \delta^i d\mathbf{v}_{[0]}
$$

(2.40)

$$
= \alpha(\mathbf{x}, \mathbf{v}, \delta) d\mathbf{x}_{[0]} + \beta(\mathbf{x}, \delta) d\mathbf{v}_{[0]}
$$

so that the inverse feedback is

$$
d\mathbf{v}_{[0]} = \hat{\alpha}(\mathbf{x}, \mathbf{u}, \delta) d\mathbf{x}_{[0]} + \hat{\beta}(\mathbf{x}, \delta) d\mathbf{u}_{[0]}.
$$

As the matrix $\beta(\mathbf{x}, \delta)$ is unimodular, one gets $\hat{\beta}(\mathbf{x}, \delta) = \beta^{-1}(\mathbf{x}, \delta)$. Accordingly, the differential representation of the closed-loop system, given by the dynamics

$$
d\dot{\mathbf{x}}_{[0]} = f(\mathbf{x}, \mathbf{u}, \delta) d\mathbf{x}_{[0]} + g(\mathbf{x}, \delta) d\mathbf{u}_{[0]}
$$

with the feedback $d\mathbf{u}_{[0]} = \alpha(\mathbf{x}, \mathbf{v}, \delta) d\mathbf{x}_{[0]} + \beta(\mathbf{x}, \delta) d\mathbf{v}_{[0]}$, reads

$$
d\dot{\mathbf{x}}_{[0]} = \hat{f}(\mathbf{x}, \mathbf{v}, \delta) d\mathbf{x}_{[0]} + \hat{g}(\mathbf{x}, \delta) d\mathbf{v}_{[0]}
$$

with

$$
\hat{f}(\mathbf{x}, \mathbf{v}, \delta) = f(\mathbf{x}, \mathbf{u}, \delta)|_{u=\alpha(\cdot)+\beta(\cdot)v} + g(\mathbf{x}, \delta)\alpha(\mathbf{x}, \mathbf{v}, \delta),
$$
$$
\hat{g}(\mathbf{x}, \delta) = g(\mathbf{x}, \delta)\beta(\mathbf{x}, \delta).
$$

Example 2.8 The feedback

$$
u_1(t) = v_1(t)
$$
$$
u_2(t) = x^2(t-1)v_1(t-2) + v_2(t)
$$

involves delays and obviously is invertible since

$$
v_1(t) = u_1(t)
$$
$$
v_2(t) = -x^2(t-1)u_1(t-2) + u_2(t).
$$

2.8   Problems

1. Consider the dynamics

$$\dot{x}(t) = \begin{pmatrix} x_1(t)x_2(t-D) + x_2(t) \\ x_1(t) + x_2(t)x_2(t-D) \end{pmatrix} + \begin{pmatrix} x_1(t) \\ 1 \end{pmatrix} u(t-2D)$$

$$y(t) = x_1(t) + x_2(t-D)$$

with D as constant delay. Find the associated differential representation.

2. Given the submodule elements $r_1(\mathbf{x}, \delta) = \begin{pmatrix} \delta \\ x_1(t-1) \\ x_2^2(t) \end{pmatrix}$,

$r_2(\mathbf{x}, \delta) = \begin{pmatrix} x_2(t) \\ x_1(t-2)\delta \\ 1 \end{pmatrix}$, compute the Polynomial Lie Bracket

$$[\mathbf{R}_1(\mathbf{x}, \epsilon), r_2(\mathbf{x}, \delta)].$$

3. Given $r_1(\mathbf{x}, \delta) = \begin{pmatrix} \delta \\ x_1(t-1) \\ x_2^2(t) \end{pmatrix}$, $r_2(\mathbf{x}, \delta) = \begin{pmatrix} x_2(t) \\ x_1(t-2)\delta \\ 1 \end{pmatrix}$, compute the Generalized Lie derivative $L_{r_1^{-1}(\mathbf{x})} r_2^0(\mathbf{x})$.

4. Verify Properties 2.1, 2.2 and 2.3 in Sect. 2.5

$$r_1(\mathbf{x}, \delta) = \begin{pmatrix} \delta^2 \\ x_2(t) \\ x_1(t-2) \end{pmatrix}, \quad r_2(\mathbf{x}, \delta) = \begin{pmatrix} x_2(t) \\ x_1(t-2)\delta^2 \\ \delta. \end{pmatrix}$$

5. Prove Property 2.1 and Property 2.2 in Sect. 2.5.

6. Consider the (linear) delay-dependent input transformation

$$u_1(t) = v_1(t) - v_2(t-1)$$
$$u_2(t) = v_1(t-1) + v_2(t) - v_2(t-2).$$

Is this transformation invertible? If yes, then write $v_1(t)$ and $v_2(t)$ in terms of $u_1(t)$, $u_2(t)$ and their delays.

7. Consider the delay-dependent state feedback

$$u_1(t) = x_1^2(t) + x_2(t-1)v_1(t-1) + v_2(t)$$
$$u_2(t) = x_2^3(t-1) + v_1(t).$$

Is this state feedback invertible? If yes, is the inverse state feedback causal?

Chapter 3
The Geometric Framework—Results on Integrability

In this chapter, we will focus our attention on the solvability of a set of partial differential equations of the first order, or equivalently the integrability problem of a set of one forms when the given variables are affected by a constant delay. As it is well known, in the delay-free case such a problem is addressed by using Frobenius theorem, and the necessary and sufficient conditions for integrability can be stated equivalently by referring to involutive distributions or involutive codistributions. Frobenius theorem is used quite frequently in the nonlinear delay-free context because it is at the basis of the solution of many control problems. This is why it is fundamental to understand how it works in the delay context.

When dealing with time-delay systems, in fact, things become more involved. A first attempt to deal with the problem can be found in Márquez-Martínez (2000) where it is shown that for a single one-form one gets results which are similar to the delay-free case, while these results cannot be extended to the general context.

As shown in Chap. 1, a first important characteristics of one-forms in the time-delay context, is that they have to be viewed as elements of a module over a certain non-commutative polynomial ring.

In the present chapter, it will also be shown that two notions of integrability have to be defined in the delay context, strong and weak integrability, which instead coincide in the delay-free case. As it will be pointed out, these main differences are linked to the notion of closure of a submodule introduced in Conte and Perdon (1995) for linear systems over rings and recalled in Chap. 1. Finally, it will also be shown that the concept of involutivity can be appropriately extended to this context through the use of the Polynomial Lie Bracket.

© The Author(s), under exclusive license to Springer Nature Switzerland AG 2021
C. Califano and C. H. Moog, *Nonlinear Time-Delay Systems*,
SpringerBriefs in Control, Automation and Robotics,
https://doi.org/10.1007/978-3-030-72026-1_3

3.1   Some Remarks on Left and Right Integrability

Let us first underline what is meant by integrability of a left- or right-submodule:

1. Integrating a given left-submodule

$$\Omega(\delta) = \text{span}_{\mathcal{K}^*(\delta]}\{\omega_1(\mathbf{x}, \delta), \ldots, \omega_k(\mathbf{x}, \delta)\}$$

generated by k one-forms independent over $\mathcal{K}^*(\delta]$ consists in finding k independent functions $\varphi_1, \ldots \varphi_k$ such that

$$\text{span}_{\mathcal{K}^*(\delta]}\{\omega_1, \ldots, \omega_k\} \subseteq \text{span}_{\mathcal{K}^*(\delta]}\{d\varphi_1, \ldots, d\varphi_k\}. \qquad (3.1)$$

2. Integrating a given right-submodule

$$\Delta(\delta] = \text{span}_{\mathcal{K}^*(\delta]}\{r_1(\mathbf{x}, \delta), \ldots, r_j(\mathbf{x}, \delta)\},$$

generated by j independent elements over $\mathcal{K}^*(\delta]$, consists in the computation of a set of $n - j$ exact differentials $d\lambda_\mu(\mathbf{x}) = \Lambda_\mu(\mathbf{x}, \delta)d\mathbf{x}_{[0]}(p)$ independent over $\mathcal{K}^*(\delta]$, which define a basis for the left-kernel of $\Delta(\delta]$.

Already these definitions enlighten two important differences with respect to the delay-free case:

First, Eq. (3.1) states that integrability of $\Omega(\delta]$ is equivalent to finding k independent functions φ_i $i \in [1, k]$ such that

$$\begin{pmatrix} \omega_1 \\ \vdots \\ \omega_k \end{pmatrix} = A(\mathbf{x}, \delta) \begin{pmatrix} d\varphi_1 \\ \vdots \\ d\varphi_k \end{pmatrix}.$$

If the matrix $A(\mathbf{x}, \delta)$ is unimodular then this means that

$$\text{span}_{\mathcal{K}^*(\delta]}\{\omega_1, \ldots, \omega_k\} \equiv \text{span}_{\mathcal{K}^*(\delta]}\{d\varphi_1, \ldots, d\varphi_k\},$$

so that the differentials $d\varphi_j$ can be expressed in terms of the one-forms ω_i's, and this is exactly what happens in the delay-free case. We will talk in this case of **Strong Integrability**. If instead the matrix $A(\mathbf{x}, \delta)$ is not unimodular then we cannot express the $d\varphi_j$'s in terms of the one-forms ω_i, since

$$\text{span}_{\mathcal{K}^*(\delta]}\{\omega_1, \ldots, \omega_k\} \subset \text{span}_{\mathcal{K}^*(\delta]}\{d\varphi_1, \ldots, d\varphi_k\}.$$

We will talk in this case of **Weak Integrability** which is then peculiar of the delay context only and is directly linked to the concept of closure of a submodule.

Secondly, such a difference does not come out if one works on the right-submodule $\Delta(\delta]$, since its left-annihilator is always closed, so that one will always find out that its left-annihilator is strongly integrable. Instead, as it will be clarified later on, when talking of integrability of a right-submodule a new notion of p-integrability needs to be introduced which characterizes the index p such that $d\lambda = \Lambda(\mathbf{x}, \delta)d\mathbf{x}_{[0]}(p)$ satisfies $\Lambda(\mathbf{x}, \delta)\Delta(\delta] = 0$.

Example 3.1 The one-form $\omega_1 = dx(t) + x(t-1)dx(t-1)$ can be written in the two following forms:

$$\omega_1 = (1 + x(t-1)\delta)dx(t) \tag{3.2}$$

and

$$\omega_1 = d(x(t) + 1/2x(t-1)^2). \tag{3.3}$$

Equation (3.2) suggests that the given one form is just weakly integrable; however, it is even strongly integrable from (3.3).

Instead, the one-form $\omega_2 = dx_1(t) + x_2(t)dx_1(t-1) = (1 + x_2(t)\delta)dx_1(t)$ is weakly integrable, but not strongly integrable, because the polynomial $1 + x_2(t)\delta$ is not invertible.

3.2 Integrability of a Right-Submodule

The following section is devoted to analyze in more detail the concept of integrability of a right-submodule. As it has been underlined at the beginning of this chapter, differently to the delay-free case, in this context it will be necessary to introduce a more general definition of integrability, namely, p-integrability. Another concept strictly linked to the integrability problem is that of involutivity. Such a notion is also introduced in this context, though it will be shown that it is not a straightforward generalization to the delay context of the standard definitions. The main features of delay systems, in fact, fully characterize these topics.

Let us now consider the right-submodule

$$\Delta(\delta] = \text{span}_{\mathcal{K}^*(\delta]}\{r_1(\mathbf{x}, \delta), \dots, r_j(\mathbf{x}, \delta)\} \tag{3.4}$$

of rank j, with the polynomial vector $r_i(\mathbf{x}, \delta) = \sum_{\ell=0}^{\bar{s}}(r_i^\ell(\mathbf{x}))^T \frac{\partial}{\partial \mathbf{x}_{[p]}}\delta^\ell \in \mathcal{K}^{*n}(\delta]$. By assumption $r_i^{\bar{s}+\ell} = 0, \forall \ell > 0$; by convention $r_i^{-k} = 0, \forall k > 0$.

As we have already underlined, integrating $\Delta(\delta]$ consists in the computation of a set of $n - j$ exact differentials $d\lambda_\mu(\mathbf{x}) = \Lambda_\mu(\mathbf{x}, \delta)d\mathbf{x}_{[0]}(p)$ independent over $\mathcal{K}^*(\delta]$, which define a basis for the left-kernel of $\Delta(\delta]$.

As it is immediately evident, one key point stands in the computation of the correct p. We will thus talk of p-integrability of a right-submodule, which is defined as follows.

Definition 3.1 (*p-integrability of a Right-Submodule*) The right-submodule

$$\Delta(\delta) = \text{span}_{\mathcal{K}^*(\delta)}\{r_1(\mathbf{x}, \delta), \ldots, r_j(\mathbf{x}, \delta)\}$$

of rank j is said to be p-integrable if there exist $n - j$ independent exact differentials

$$d\lambda_\mu(\mathbf{x}) = \Lambda_\mu(\mathbf{x}, \delta)d\mathbf{x}_{[0]}(p), \ \mu \in [1, n - j]$$

such that the $d\lambda_\mu(\mathbf{x})$'s lay in the left-kernel of $\Delta(\delta)$, that is, $d\lambda_\mu(\mathbf{x})r_i(\mathbf{x}, \delta) = 0$, for $i \in [1, j]$ and $\mu \in [1, n - j]$, and any other exact differential $d\bar{\lambda}(\mathbf{x}) \in \Delta^\perp(\delta)$ can be expressed as a linear combination over $\mathcal{K}^*(\delta)$ of such $d\lambda_\mu(\mathbf{x})$'s.

Definition 3.2 (*integrability of a Right-Submodule*) The right-submodule $\Delta(\delta)$ of rank j, given by (3.4), is said to be integrable if there exists some finite integer p such that $\Delta(\delta)$ is p-integrable.

Example 3.2 Consider, for instance,

$$\Delta(\delta) = \text{span}_{\mathcal{K}^*(\delta)}\left\{\begin{pmatrix} -x_1(2)\delta \\ x_2(2) \end{pmatrix}\right\}.$$

According to the above definition, $\Delta(\delta)$ is 2-integrable, since

$$d\lambda = d(x_1(2)x_2(1)) = (x_2(1), x_1(2)\delta)dx(2) \perp \Delta(\delta).$$

How to check the existence of such a solution and how to compute it is the topic of the present chapter.

To this end, starting from the definitions of Generalized Lie Bracket, Lie Bracket, and Polynomial Lie Bracket given in Chap. 2, the notions of Involutivity and Involutive Closure of a right-submodule are introduced next. They represent the nontrivial generalization of the standard definitions used in the delay-free context, which can be recovered as a special case. These definitions play a fundamental role in the integrability conditions.

3.2.1 Involutivity of a Right-Submodule Versus its Integrability

As already recalled, for right-submodules to deal with integrability, the involutivity concept must be defined.

Definition 3.3 (*Involutivity*) Consider the right-submodule

$$\Delta(\delta) = \text{span}_{\mathcal{K}^*(\delta]} \{r_1(\mathbf{x}, \delta), \ldots, r_j(\mathbf{x}, \delta)\}$$

of rank j, with $r_i(\mathbf{x}, \delta) = \sum_{l=0}^{s} r_i^l(\mathbf{x}_{[s_i, s]})\delta^l$ and let $\Delta_c(\delta)$ be its right closure. Then $\Delta(\delta)$ is said to be involutive if for any pair of indices $i, \ell \in [1, j]$ the Lie Bracket $[r_i(\mathbf{x}, \delta), r_\ell(\mathbf{x}, \delta)]$ satisfies

$$\text{span}_{\mathcal{K}^*(\delta]}\{[r_i(\mathbf{x}, \delta), r_\ell(\mathbf{x}, \delta)]\} \in \Delta_c(\delta). \tag{3.5}$$

Remark 3.1 Definition 3.3 includes as a special case the notion of involutivity of a distribution. The main feature is that starting from a given right-submodule, its involutivity implies that the vectors obtained through the Lie Bracket of two elements of the submodule cannot be obtained as a linear combination of the generators of the given submodule, but it is a linear combination of the generators of its right closure. For finite dimensional delay-free systems, distributions are closed by definition, so there is no such a difference.

The definition of involutivity of a submodule is crucial for the integrability problem, as enlightened in the next theorem.

Theorem 3.1 *The right-submodule*

$$\Delta(\delta) = \text{span}_{\mathcal{K}^*(\delta]} \{r_1(\mathbf{x}, \delta), \ldots, r_j(\mathbf{x}, \delta)\}$$

of rank j is 0-integrable if and only if it is involutive and its left-annihilator is causal.

Hereafter, the proof is reported in order to make the reader familiar with the Polynomial Lie Bracket. The necessity part simply shows that if the left-annihilator of $\Delta(\delta)$ is integrable, then necessarily all the Polynomial Lie Brackets of the vectors in $\Delta(\delta)$ must be necessarily in $\Delta(\delta)$. The sufficiency part, instead, starts by associating a finite involutive distribution to $\Delta(\delta)$, and then showing that its integrability allows us to compute the exact one forms that span the left-annihilator of $\Delta(\delta)$.

Proof Necessity. Assume that there exist $n - j$ causal exact differentials $d\lambda_i(\mathbf{x}) = \Lambda_i(\mathbf{x}, \delta)d\mathbf{x}_{[0]}$, independent over $\mathcal{K}^*(\delta]$ which are in $\Delta^\perp(\delta)$. Let ρ denote the maximum between the delay in the state variable and the degree in δ. Then

$$\Lambda_\mu(\mathbf{x}_{[\rho]}, \delta)r_\ell(\mathbf{x}, \delta) = 0, \ \forall \mu \in [1, n - j], \ \forall \ell \in [1, j]. \tag{3.6}$$

The time derivative of (3.6) along $\mathbf{R}_q(\mathbf{x}_{[s_1, s]}, \epsilon)$ yields $\forall \mu \in [1, n - j], \forall \ell \in [1, j]$

$$\dot{\Lambda}_\mu(\mathbf{x}, \delta)|_{\dot{\mathbf{x}}_{[0]} = \mathbf{R}_q(\mathbf{x}, \epsilon)}r_\ell(\mathbf{x}, \delta) + \Lambda_\mu(\mathbf{x}, \delta)\dot{r}_\ell(\mathbf{x}, \delta)|_{\dot{\mathbf{x}}_{[0]} = \mathbf{R}_q(\mathbf{x}, \epsilon)} = 0.$$

Multiplying on the right by δ^{s_1}, one gets

$$\sum_{i,k=0}^{\rho} \frac{\partial \Lambda_\mu^i(\mathbf{x})}{\partial \mathbf{x}(-k)}\mathbf{R}_q(\mathbf{x}(-k), \epsilon(-k))\delta^i r_\ell(\mathbf{x}, \delta)\delta^{s_1} + \Lambda_\mu(\mathbf{x}, \delta)\dot{r}_\ell(\mathbf{x}, \delta)|_{\dot{\mathbf{x}}_{[0]} = \mathbf{R}_q(\mathbf{x}, \epsilon)}\delta^{s_1} = 0,$$

that is, recalling that $\dfrac{\partial \Lambda^i_\mu(\mathbf{x})}{\partial \mathbf{x}(-k)} = \dfrac{\partial \Lambda^k_\mu(\mathbf{x})}{\partial \mathbf{x}(-i)}$,

$$\sum_{i,k=0}^{\rho} \frac{\partial \Lambda^k_\mu(\mathbf{x})}{\partial \mathbf{x}(-i)} \mathbf{R}_q(\mathbf{x}(-k), \epsilon(-k)) \delta^i r_\ell(\mathbf{x}, \delta) \delta^{s_1} +$$

$$+ \Lambda_\mu(\mathbf{x}, \delta) \sum_{k=0}^{s+s_1} \frac{\partial \mathbf{R}_q(\mathbf{x}, \epsilon)}{\partial \mathbf{x}(s_1 - k)} \delta^k r_\ell(\mathbf{x}(s_1), \delta) = -\Lambda_\mu(\mathbf{x}, \delta)[\mathbf{R}_q(\mathbf{x}, \epsilon), r_\ell(\mathbf{x}, \delta]. \quad (3.7)$$

Moreover, since $\lambda_\mu(\mathbf{x})$ is causal then $\dfrac{\partial \Lambda^k_\mu(\mathbf{x})}{\partial \mathbf{x}(s_1 - i)} = 0$ for $i \in [0, s_1 - 1]$; since $\Lambda_\mu(\mathbf{x}, \delta) r_q(\mathbf{x}, \delta) = 0$, then also $\sum_{k=0}^{s} \Lambda^k_\mu(\mathbf{x}) \mathbf{R}_q(\mathbf{x}(-k), \epsilon(-k)) = 0$, so that for $i \in [0, s + s_1]$,

$$\sum_{k=0}^{s} \frac{\partial \Lambda^k_\mu(\mathbf{x})}{\partial \mathbf{x}(-i)} \mathbf{R}_q(\mathbf{x}(-k), \epsilon(-k)) + \sum_{k=0}^{s} \Lambda^k_\mu(\mathbf{x}) \frac{\partial \mathbf{R}_q(\mathbf{x}(-k), \epsilon(-k))}{\partial \mathbf{x}(-i)} = 0.$$

It follows, through standard computations, that

$$\sum_{i=0}^{s+s_1} \sum_{k=0}^{s} \frac{\partial \Lambda^k_\mu(\mathbf{x})}{\partial \mathbf{x}(s_1 - i)} \mathbf{R}_q(\mathbf{x}(-k), \epsilon(-k)) \delta^i = -\sum_{i=0}^{s+s_1} \Lambda_\mu(\mathbf{x}, \delta) \frac{\partial \mathbf{R}_q(\mathbf{x}, \epsilon)}{\partial \mathbf{x}(s_1 - i)},$$

which, substituted in (3.7), leads to

$$\Lambda_\mu(\mathbf{x}, \delta)[\mathbf{R}_q(\mathbf{x}, \epsilon), r_\ell(\mathbf{x}, \delta)] = 0, \quad \forall \epsilon.$$

Since the previous relation must be satisfied $\forall \mu \in [1, n - j]$, and $\forall \ell, q \in [1, j]$, and recalling the link between the Polynomial Lie Bracket and the Lie Bracket highlightened in Eq. (2.24), then necessarily $\Delta(\delta]$ must be involutive.

Sufficiency. Let $\omega(\mathbf{x}_{[\hat{s}]}, \delta) = (\omega^T_1(\mathbf{x}_{[\hat{s}]}, \delta), \dots, \omega^T_{n-j}(\mathbf{x}_{[\hat{s}]}, \delta))^T$ be the left-annihilator of $(r_1(\mathbf{x}_{[s_1, s]}, \delta), \dots, r_j(\mathbf{x}_{[s_k, s]}, \delta))$. Let $\bar{s} = max\{s_1, \dots, s_k\}$ and $\rho = max\{\hat{s}, deg(\omega(\mathbf{x}, \delta))\}$, that is, for $k \in [1, n - j]$, $\omega_k(\mathbf{x}, \delta) = \sum_{\ell=0}^{\rho} \omega^\ell_k(\mathbf{x}_{[\rho]}) \delta^\ell$. Set $\Omega = (0, \dots, 0, \omega^0(\mathbf{x}_{[\rho]}), \dots, \omega^\rho(\mathbf{x}_{[\rho]}), 0, \dots, 0)$, where ω^0 is preceded by $\bar{s}$ 0-blocks and set $\Delta_i := \Delta_{[\bar{s}, i+s]} \subset span\{\frac{\partial}{\partial \mathbf{x}_{[0]}(\bar{s})}, \dots, \frac{\partial}{\partial \mathbf{x}_{[0]}(-i-s)}\}$ as

$$\Delta_i = span_{K^*} \begin{pmatrix} I_{n\bar{s}} & * & * & 0 & \cdots & & & 0 \\ 0 & \mathbf{r}^0(\mathbf{x}) & \cdots & \mathbf{r}^\ell(\mathbf{x}) & 0 & & & \\ \vdots & 0 & \ddots & & & \ddots & \ddots & \\ \vdots & & 0 & \mathbf{r}^0(\mathbf{x}(-i)) & \cdots & \mathbf{r}^\ell(\mathbf{x}(-i)) & 0 \\ 0 & 0 & \cdots & 0 & \cdots & 0 & I_{ns} \end{pmatrix}. \quad (3.8)$$

$$\underbrace{}_{\Delta_{i0}}$$

By assumption $\omega(\mathbf{x}, \delta)$ is causal and for any two vector fields $\tau_\ell \in \Delta_{i0}$, $\ell = 1, 2$, $i \geq \rho$, $\Omega \tau_\ell = 0$, and $\Omega[\tau_1, \tau_2] = 0$. Moreover, since $i \geq \rho$, $\Omega \frac{\partial}{\partial x_\ell(-i-p)} = 0$, $\forall \ell \in [1, n]$, $\forall p \in [1, s]$. It follows that $\Omega[\tau_1, \frac{\partial}{\partial x_\ell(-i-p)}] = 0$, since $\frac{\partial(\Omega \tau_1)}{\partial x_\ell(-i-p)} = \Omega \frac{\partial \tau_1}{\partial x_\ell(-i-p)} = 0$. Analogously, since Ω is causal, then for any $p \in [1, \bar{s}]$, $\frac{\partial(\Omega \tau_1)}{\partial x_\ell(+p)} = \Omega \frac{\partial \tau_1}{\partial x_\ell(+p)} = 0$, which shows that $\Omega[\tau_1, \frac{\partial}{\partial x_\ell(+p)}] = 0$, so that $\Omega \perp \bar{\Delta}_i$. As a consequence, there exist at least $n - j$ causal exact differentials, independent over $\mathcal{K}^*$ which lay in the left-annihilator of $\bar{\Delta}_i$. It remains to show that there are also $n - j$ causal exact differentials, independent over $\mathcal{K}^*(\delta)$, which lay in the left-annihilator of $\Delta(\delta)$. This follows immediately by noting that if $d\lambda_1, \ldots, d\lambda_\mu$, $\mu \leq n - j$ is a basis for $\Delta^\perp(\delta)$, since Ω is 0-integrable then $\omega(\mathbf{x}, \delta)d\mathbf{x}_{[0]} = \sum_{i=1}^{\mu} \alpha_i(\mathbf{x}, \delta)d\lambda_i$. Since the $\omega_i(\mathbf{x}, \delta)d\mathbf{x}_{[0]}$'s are $n - j$ and by assumption they are independent over $\mathcal{K}^*(\delta)$, then necessarily $\mu = n - j$.

A direct consequence of the proof of Theorem 3.1 is the definition of an upper bound on the maximum delay appearing in the exact differentials which generate a basis for the left-annihilator of $\Delta(\delta)$. This is pointed out in the next corollary.

Corollary 3.1 *Let the right-submodule*

$$\Delta(\delta) = \mathrm{span}_{\mathcal{K}^*(\delta)} \{r_1(\mathbf{x}, \delta), \ldots, r_j(\mathbf{x}, \delta)\}$$

of rank j, with $r_i(\mathbf{x}, \delta) = \sum_{l=0}^{\bar{s}} r_i^l(\mathbf{x}_{[s_i, s]})\delta^l$, be 0-integrable. Then the maximum delay which characterizes the exact differentials generating the left-annihilator of $\Delta(\delta)$ is not greater than $j\bar{s} + s$.

Proof The proof of Theorem 3.1 shows that if ρ is the maximum between the degree in δ and the largest delay affecting the state variables in the left-annihilator $\Omega(\mathbf{x}_{[\bar{p}]}, \delta)$ of $\Delta(\delta)$, then the exact differentials are affected by a maximum delay which is not greater than ρ. On the other hand, according to Lemma 1.1 in Chap. 1, $deg(\Omega(\mathbf{x}, \delta)) \leq j\bar{s}$, whereas $\bar{p} \leq s + j\bar{s}$, which shows that $\rho \leq j\bar{s} + s$.

The result stated by Theorem 3.1 which is itself an important achievement plays also a key role in proving a series of fundamental results which are enlightened hereafter.

3.2.2 Smallest 0-Integrable Right-Submodule Containing $\Delta(\delta)$

If the given submodule $\Delta(\delta)$ is not 0-integrable, one may be interested in computing the smallest 0-integrable right-submodule containing it. This in turn will allow to identify the maximum number of independent exact one-forms which stand in the left-annihilator of $\Delta(\delta)$. The following definition needs to be introduced, which generalizes the notion of involutive closure of a distribution to the present context.

Definition 3.4 (*Involutive closure*) Given the right-submodule

$$\Delta(\delta] = \text{span}_{\mathcal{K}^*(\delta]} \left\{ r_1(\mathbf{x}, \delta), \ldots, r_j(\mathbf{x}, \delta) \right\}$$

of rank j, with $r_i(\mathbf{x}, \delta) = \sum_{l=0}^{s} r_i^l(\mathbf{x}_{[s_i, s]}) \delta^l$, let $\Delta_c(\delta]$ be its right closure. Then its involutive closure $\bar{\Delta}(\delta]$ is the smallest submodule, which contains $\Delta_c(\delta]$ and which is involutive.

Accordingly, the following result can be stated.

Theorem 3.2 *Consider the right-submodule*

$$\Delta(\delta] = \text{span}_{\mathcal{K}^*(\delta]} \left\{ r_1(\mathbf{x}, \delta), \ldots, r_j(\mathbf{x}, \delta) \right\}$$

of rank j and let $\bar{\Delta}(\delta]$ be its involutive closure and assume that the left-annihilator of $\bar{\Delta}(\delta]$ is causal. Then $\bar{\Delta}(\delta]$ is the smallest 0-integrable right-submodule containing $\Delta(\delta]$.

Example 3.3 Let us consider

$$\Delta(\delta] = \text{span}_{\mathcal{K}^*(\delta]} \left\{ \begin{pmatrix} x_1 x_1(1) \delta^2 \\ -x_2 x_1(2) \delta - x_1 \delta^2 \end{pmatrix} \right\}.$$

$\Delta(\delta]$ is not closed. Clearly, its right closure is given by

$$\Delta_c(\delta] = \left\{ \begin{pmatrix} x_1 x_1(1) \delta \\ -x_2 x_1(2) - x_1 \delta \end{pmatrix} \right\}.$$

To check the involutivity, we set

$$r(\mathbf{x}, \delta) = \begin{pmatrix} x_1 x_1(1) \delta \\ -x_2 x_1(2) - x_1 \delta \end{pmatrix}, \quad \mathbf{R}(\mathbf{x}, \epsilon) = \begin{pmatrix} x_1 x_1(1) \epsilon(-1) \\ -x_2 x_1(2) \epsilon(0) - x_1 \epsilon(-1) \end{pmatrix}.$$

We then compute the Polynomial Lie Bracket

$$[\mathbf{R}(\mathbf{x}, \epsilon), r(\mathbf{x}, \delta)] = \dot{r}(\mathbf{x}, \delta)|_{\dot{\mathbf{x}}_{[0]} = \mathbf{R}(\mathbf{x}, \epsilon)} \delta^2 - \sum_{\ell=0}^{2} \frac{\partial \mathbf{R}(\mathbf{x}, \epsilon)}{\partial x(2 - \ell)} \delta^\ell r(\mathbf{x}(2), \delta)$$

$$= \begin{pmatrix} x_1 x_1^2(1) \epsilon(-1) \delta + x_1 x_1(1) x_1(2) \epsilon(0) \delta \\ x_1(2)(x_2 x_1(2) \epsilon(0) - x_1 \epsilon(-1)) - x_2 x_1(2) x_1(3) \epsilon(1) - x_1 x_1(1) \epsilon(-1) \delta \end{pmatrix} \delta^2$$

$$- \begin{pmatrix} x_1(1) \epsilon(-1) \delta x_1(2) x_1(3) \delta + x_1 \epsilon(-1) \delta^2 x_1(2) x_1(3) \delta \\ -x_2 \epsilon(0) x_1(2) x_1(3) \delta + \epsilon(0) x_1(2) \delta^2 (x_2(2) x_1(4) - x_1(2) \delta) - \epsilon(-1) \delta (x_1(2) x_1(3) \delta) \end{pmatrix}$$

$$= \begin{pmatrix} x_1 x_1(1) \delta \\ -x_2 x_1(2) - x_1 \delta \end{pmatrix} x_1(3) \left(\delta^2 \epsilon(3) - \delta \epsilon(1) \right)$$

$$= r(\mathbf{x}, \delta) x_1(3) \left(\delta^2 \epsilon(3) - \delta \epsilon(1) \right).$$

Thus, $\Delta_c(\delta]$ is involutive and is the smallest 0-integrable right-submodule containing $\Delta(\delta]$.

3.2.3 p-Integrability

The approach presented in this book allows us to state a more general result concerning p-integrability which is independent of any control system. This is done hereafter.

Theorem 3.3 *The right-submodule*

$$\Delta(\delta] = \text{span}_{\mathcal{K}^*(\delta]} \{r_1(\mathbf{x}, \delta), \ldots, r_j(\mathbf{x}, \delta)\}$$

of rank j is p-integrable if and only if

$$\hat{\Delta}(\delta] = \Delta(\mathbf{x}(-p), \delta) = \text{span}_{\mathcal{K}^*(\delta]} \{r_1(\mathbf{x}(-p), \delta), \ldots, r_j(\mathbf{x}(-p), \delta)\}$$

is 0-integrable.

Proof Assume that $\Delta(\delta]$ is p-integrable. Then there exist $n - j$ independent exact differentials $d\lambda_i(\mathbf{x}) = \Lambda_i(\mathbf{x}, \delta)d\mathbf{x}_{[0]}(p)$ such that, denoting by $\Lambda(\mathbf{x}, \delta) = (\Lambda_1^T(\mathbf{x}, \delta), \ldots, \Lambda_{n-j}^T(\mathbf{x}, \delta))^T$, then $\Lambda(\mathbf{x}, \delta)\Delta(\delta] = 0$. Consequently, for $i \in [1, j]$,

$$\delta^p \Lambda(\mathbf{x}, \delta)r_i(\mathbf{x}, \delta) = \Lambda(\mathbf{x}(-p), \delta)r_i(\mathbf{x}(-p), \delta)\delta^p = 0,$$

that is, $\Lambda(\mathbf{x}(-p), \delta)\hat{\Delta}(\delta] = 0$. On the other hand,

$$\delta^p \Lambda(\mathbf{x}, \delta)d\mathbf{x}_{[0]}(p) = \Lambda(\mathbf{x}(-p), \delta)d\mathbf{x}_{[0]},$$

which thus proves that $\hat{\Delta}(\delta]$ is 0-integrable. Of course, if $\hat{\Delta}(\delta]$ is 0-integrable, there exist $n - j$ exact differentials $d\bar{\lambda}_i(\mathbf{x}) = \bar{\Lambda}_i(\mathbf{x}, \delta)d\mathbf{x}_{[0]}$ such that $\bar{\Lambda}(\mathbf{x}, \delta)\hat{\Delta}(\delta] = 0$. As a consequence also $\bar{\Lambda}(\mathbf{x}, \delta)\hat{\Delta}(\delta]\delta^p = 0$, which shows that $\hat{\Delta}(\mathbf{x}(p), \delta) = \Delta(\delta]$ is p-integrable.

Example 3.4 Let us consider again the submodule

$$\Delta(\delta] = \text{span}_{\mathcal{K}^*(\delta]} \left\{ \begin{pmatrix} -x_1(2)\delta \\ x_2(2) \end{pmatrix} \right\}$$

of Example 3.2. We first check if it is 0-integrable. To this end, we set

$$\mathbf{R}(\mathbf{x}, \epsilon) = \left\{ \begin{pmatrix} -x_1(2)\epsilon(-1) \\ x_2(2)\epsilon(0) \end{pmatrix} \right\}$$

and compute the Polynomial Lie Bracket

$$[\mathbf{R}(\mathbf{x}, \epsilon), r(\mathbf{x}, \delta)] = \dot{r}(\mathbf{x}, \delta)|_{\dot{\mathbf{x}}_{[0]}=\mathbf{R}(\mathbf{x},\epsilon)}\delta^2 - \frac{\partial \mathbf{R}(\mathbf{x}, \epsilon)}{\partial x(2)} r(\mathbf{x}(2), \delta)$$

$$= \begin{pmatrix} x_1(4)\epsilon(1)\delta \\ x_2(4)\epsilon(2) \end{pmatrix} \delta^2 - \begin{pmatrix} \epsilon(-1)x_1(4)\delta \\ \epsilon(0)x_2(4) \end{pmatrix}$$

$$= \begin{pmatrix} x_1(4)\delta \\ x_2(4) \end{pmatrix} \left(\delta^2 \epsilon(4) - \epsilon(0) \right)$$

which is not in $\Delta(\delta]$. To check 2-integrability, we have to consider

$$\hat{\Delta}(\delta] = \Delta(\mathbf{x}(-2), \delta) = \text{span}_{\mathcal{K}^*(\delta]} \left\{ \begin{pmatrix} -x_1\delta \\ x_2 \end{pmatrix} \right\}.$$

We set

$$\mathbf{R}(\mathbf{x}(-2), \epsilon) = \begin{pmatrix} -x_1\epsilon(-1) \\ x_2\epsilon(0) \end{pmatrix}$$

and compute the Polynomial Lie Bracket
$[\mathbf{R}(\mathbf{x}(-2), \epsilon), \hat{r}(\mathbf{x}(-2), \delta)] =$

$$= \dot{r}(\mathbf{x}(-2), \delta)|_{\dot{\mathbf{x}}_{[0]}=\mathbf{R}(\mathbf{x}(-2),\epsilon)} - \frac{\partial \mathbf{R}(\mathbf{x}(-2), \epsilon)}{\partial x(0)} r(\mathbf{x}(-2), \delta)$$

$$= \begin{pmatrix} x_1(0)\epsilon(-1)\delta \\ x_2(0)\epsilon(0) \end{pmatrix} - \begin{pmatrix} \epsilon(-1)x_1(0)\delta \\ \epsilon(0)x_2(0) \end{pmatrix} = 0$$

which shows that $\Delta(\delta]$ is 2-integrable. In fact, one gets that $d\lambda = d(x_1(2)x_2(1)) = (x_2(1), x_1(2))\delta)dx(2))$ is in the left-kernel of $\Delta(\delta]$.

3.2.4 Bicausal Change of Coordinates

A major problem in control theory stands in the possibility of describing the given system in some different coordinates which may put in evidence particular structural properties. As already underlined in Chap. 1, in the delay context, it is fundamental to be able to compute bicausal change of coordinates, that is, diffeomorphisms which are causal and admit a causal inverse as defined by Definition 1.1.

As it will be shown in this section, the previous results on the 0-integrability of a right-submodule have important outcomes also in the definition of a bicausal change of coordinates. Lemma 3.1 is instrumental for determining whether or not a given set of one-forms can be used to obtain a bicausal change of coordinates.

Lemma 3.1 *Given* $n - k$ *independent functions on* $\mathcal{R}^n$, $\lambda_i(\mathbf{x}_{[\alpha]})$, $i \in [1, n - k]$, *such that*

$$\mathrm{span}_{\mathcal{K}(\delta)}\{d\lambda_1, \ldots, d\lambda_{n-k}\}$$

is closed and its right-annihilator is causal, there exists a $d\theta_1(\mathbf{x}_{[\bar{\alpha}]})$ *independent of the* $d\lambda_i(\mathbf{x}_{[\alpha]})$*'s* $i \in [1, n - k]$ *over* $\mathcal{K}(\delta)$ *and such that*

$$\mathrm{span}_{\mathcal{K}(\delta)}\{d\lambda_1, \ldots, d\lambda_{n-k}, d\theta_1\}$$

is closed and its right-annihilator is causal.

While the proof is omitted and can be found in Califano and Moog (2017), starting from it, it is immediate to get the following result.

Theorem 3.4 *Given* k *functions* $\lambda_i(\mathbf{x}_{[\alpha]})$, $i \in [1, k]$, *whose differentials are independent over* $\mathcal{K}(\delta)$, *there exist* $n - k$ *functions* $\theta_j(\mathbf{x}_{[\bar{\alpha}]})$, $j \in [1, n - k]$ *such that*

$$\mathrm{span}_{\mathcal{K}(\delta)}\{d\lambda_1, \ldots, d\lambda_k, d\theta_1, \ldots, d\theta_{n-k}\} \equiv \mathrm{span}_{\mathcal{K}(\delta)}\{d\mathbf{x}_{[0]}\}$$

if and only if

$$\mathrm{span}_{\mathcal{K}(\delta)}\{d\lambda_1, \ldots, d\lambda_k\}$$

is closed and its right-annihilator is causal. As a consequence, $d\mathbf{z}_{[0]} = (d\lambda_1^T, \ldots, d\lambda_k^T, d\theta_1^T, \ldots, d\theta_{n-k}^T)^T$ *defines a bicausal change of coordinates.*

Proof If the k exact differentials $d\lambda_i(\mathbf{x})$ can be completed to span all $d\mathbf{x}_{[0]}$ over $\mathcal{K}(\delta)$ then necessarily $\mathrm{span}_{\mathcal{K}(\delta)}\{d\lambda_1, \ldots, d\lambda_k\}$ must be closed and its right-annihilator must be causal. On the contrary, due to Lemma 3.1, if $\mathrm{span}_{\mathcal{K}(\delta)}\{d\lambda_1, \ldots, d\lambda_k\}$ is closed and its right-annihilator is causal then one can compute an exact differential $d\theta_1$ independent over $\mathcal{K}(\delta)$ of the $d\lambda_i$'s and such that $\mathrm{span}_{\mathcal{K}(\delta)}\{d\lambda_1, \ldots, d\lambda_k, d\theta_1\}$ is closed and its right-annihilator is causal. Iterating, one gets the result.

Example 3.5 Consider the continuous-time system

$$\dot{x}_1(t) = x_2(t) + x_2(t - 1)$$
$$\dot{x}_2(t) = u(t)$$
$$\dot{x}_3(t) = x_3(t) - x_1(t - 1)$$
$$y(t) = x_1(t).$$

The output and its derivatives are given by

$$y(t) = x_1(t)$$
$$\dot{y}(t) = x_2(t) + x_2(t - 1)$$
$$\ddot{y}(t) = u(t) + u(t - 1).$$

The functions y and $\dot{y}$ could be used to define a change of coordinates if they satisfied the conditions of Lemma 3.1, that is

$$\mathcal{Y} = \text{span}_{\mathcal{K}(\delta]}\{dy, d\dot{y}\} = \text{span}_{\mathcal{K}(\delta]}\{dx_{1,[0]}, (1+\delta)dx_{2,[0]}\}$$

were closed and its right-annihilators were causal.

Unfortunately, while the right-annihilator of $\mathcal{Y}$ is causal as it is given by $\begin{pmatrix} 0 \\ 0 \\ 1 \end{pmatrix}$, $\mathcal{Y}$ itself is not closed, since it is easily verified that its closure is

$$\mathcal{Y}_c = \text{span}_{\mathcal{K}(\delta]}\{dx_{1,[0]}, dx_{2,[0]}\}.$$

As a consequence, y and $\dot{y}$ cannot be used directly to define a change of coordinates.

Example 3.6 Consider the function $\lambda = x_1(t-1)x_2(t) + x_2^2(t)$ on $\mathcal{R}^2$. In this case,

$$\begin{aligned} \mathcal{L} &= \text{span}_{Kd}\{x_2\delta dx_{1,[0]} + (x_1(-1) + 2x_2)dx_{2,[0]}\} \\ &= \text{span}_{\mathcal{K}(\delta]}\{(x_2\delta \;\; x_1(-1) + 2x_2)\, d\mathbf{x}_{[0]}\}. \end{aligned}$$

The right-annihilator is given by

$$r(\mathbf{x}, \delta) = \begin{pmatrix} -2x_2(1) - x_1 \\ x_2\delta, \end{pmatrix}$$

which is not causal. Thus, the function λ cannot be used as a basis for a change of coordinates.

3.3 Integrability of a Left-Submodule

We now address the problem of integrability working directly on one-forms. In the present paragraph, a set of one-forms $\{\omega_1, \ldots, \omega_k\}$ independent over $\mathcal{K}^*(\delta]$ is considered. Considering one-forms as elements of $\mathcal{M}$ naturally leads to two different notions of integrability. Instead, if one-forms are considered as elements of the vector space $\mathcal{E}$, there is only one single notion of integrability.

In fact, as it happens in the delay-free case, if the one-forms $\{\omega_1 \ldots, \omega_k\}$ are considered over $\mathcal{K}^*$, then they are said to be integrable if there exists an invertible matrix $A \in \mathcal{K}^{*k \times k}$ and functions $\varphi = (\varphi_1, \ldots, \varphi_k)^T$, such that $\omega = Ad\varphi$. The full rank of A guarantees the invertibility of A, since $\mathcal{K}^*$ is a field. Instead, if the one-forms $\{\omega_1 \ldots, \omega_k\}$ are viewed as elements of the module $\mathcal{M}$, then the matrix $A \in \mathcal{K}^{*k \times k}(\delta]$ instead of $\mathcal{K}^{*k \times k}$. Since $A(\delta)$ may be of full rank but not unimodular, it is necessary to distinguish two cases. Accordingly, one has the following two definitions of integrability.

Definition 3.5 A set of k one-forms $\{\omega_1, \ldots, \omega_k\}$, independent over $\mathcal{K}^*(\delta]$, is said to be strongly integrable if there exist k independent functions $\{\varphi_1, \ldots, \varphi_k\}$, such that

$$\text{span}_{\mathcal{K}^*(\delta]}\{\omega_1, \ldots, \omega_k\} = \text{span}_{\mathcal{K}^*(\delta]}\{d\varphi_1, \ldots, d\varphi_k\}.$$

A set of k one-forms $\{\omega_1, \ldots, \omega_k\}$, independent over $\mathcal{K}(\delta]$, is said to be weakly integrable if there exist k independent functions $\{\varphi_1, \ldots, \varphi_k\}$, such that

$$\text{span}_{\mathcal{K}^*(\delta]}\{\omega_1, \ldots, \omega_k\} \subseteq \text{span}_{\mathcal{K}^*(\delta]}\{d\varphi_1, \ldots, d\varphi_k\}.$$

If the one-forms $\omega = (\omega_1^T, \ldots, \omega_k^T)^T$ are strongly (respectively weakly) integrable, then the left-submodule $\text{span}_{\mathcal{K}^*(\delta]}\{\omega_1, \ldots, \omega_k\}$ is said to be strongly (respectively weakly) integrable.

Clearly, strong integrability yields weak integrability. Also, the one-forms ω are weakly integrable if and only if there exists a full rank matrix $A(\delta) \in \mathcal{K}(\delta]^{*k \times k}$ and functions $\varphi = (\varphi_1^T, \ldots, \varphi_k^T)^T$ such that $\omega = A(\delta)d\varphi$. If in addition the matrix $A(\delta)$ can be chosen to be unimodular, then the one-forms ω are also strongly integrable.

Remark 3.2 It should be noted that the integrability of a closed left-submodule $\text{span}_{\mathcal{K}(\delta]}\{\omega_1, \ldots, \omega_k\}$ always implies strong integrability. As a consequence, the two notions of strong and weak integrability coincide in case of delay-free one forms.

Integrability of a set of k one-forms $\{\omega_1, \ldots, \omega_k\}$ is tested thanks to the so-called

Derived Flag Algorithm (DFA):

Starting from a given I_0 the algorithm computes

$$I_i = \text{span}_{\mathcal{K}}\{\omega \in I_{i-1} \mid d\omega = 0 \mod I_{i-1}\}. \tag{3.9}$$

The sequence (3.9) converges as it defines a strictly decreasing sequence of vector spaces I_i and by the standard Frobenius theorem, the limit I_∞ has an exact basis, which represents the largest integrable codistribution contained in I_0.

In order to define I_0, one has to note that when considering a set of k one-forms $\{\omega_1, \ldots, \omega_k\}$, some shifts of ω_i are required for integration. It follows that the initialization

$$I_0^p = \text{span}_{\mathcal{K}}\{\omega_1, \ldots, \omega_k, \omega_1(-1), \ldots, \omega_k(-1), \ldots, \omega_1(-p), \ldots, \omega_k(-p)\} \tag{3.10}$$

allows to compute the smallest number of time shifts required for the given one-forms for the maximal integration of the submodule. More precisely, the sequence I_i^p defined by (3.9) converges to an integrable vector space

$$I_\infty^p = \text{span}_{\mathcal{K}}\{d\varphi_1^p, \ldots, d\varphi_{\gamma_p}^p\} \tag{3.11}$$

for some $\gamma_p \geq 0$. By definition, $d\varphi_i^p \in \mathrm{span}_{\mathcal{K}(\delta)}\{\omega_1, \ldots, \omega_k\}$ for $i = 1, \ldots, \gamma_p$ and $p \geq 0$. The exact one-forms $d\varphi_i^p$, $i = 1, \ldots, \gamma_p$ are independent over $\mathcal{K}$, but may not be independent over $\mathcal{K}(\delta)$. A basis for $\mathrm{span}_{\mathcal{K}(\delta)}\{d\varphi_1^p, \ldots, d\varphi_{\gamma_p}^p\}$ is obtained by computing a basis for

$$I_\infty^0 \cup \bigcup_{i=1}^{p} [I_\infty^i \mod(I_\infty^{i-1}, \delta I_\infty^{i-1})]$$

as $I_\infty^i + \delta I_\infty^i \subset I_\infty^{i+1}$.

Remark 3.3 A different initialization of the derived flag algorithm is

$$\tilde{I}_0^p = \mathrm{span}_{\mathcal{K}}\{\mathrm{span}_{\mathcal{K}(\delta)}\{\omega_1, \ldots, \omega_k\} \cap \mathrm{span}_{\mathcal{K}}\{dx(t), \ldots, dx(t-p)\}\}, \quad (3.12)$$

which allows to compute for each $p \geq 0$, the exact differentials contained in the given submodule and which depend on $x(t), \ldots, x(t-p)$ only. Both initializations allow the algorithm to converge toward the same integrable submodule over $\mathcal{K}(\delta)$, but following different steps, as shown in the next example.

Example 3.7 Let $\mathrm{span}_{\mathcal{K}(\delta)}\{dx(t-2)\}$. On one hand, the initialization (3.10) is completed for $p = 0$ as no time shift of $dx(t-2)$ is required for its integration. On the other hand, initialization (3.12) yields a 0 limit for $p = 0$ and $p = 1$ as the exact differential involves larger delays than $x(t)$ and $x(t-1)$. The final result is obtained for $p = 2$.

Assume that the maximum delay that appears in $\{\omega_1, \ldots, \omega_k\}$ (either in the coefficients or differentials) is s. The necessary and sufficient condition for strong integrability of the one-forms $\{\omega_1, \ldots, \omega_k\}$ is given by the following theorem in terms of the limit I_∞^p.

Theorem 3.5 *A set of one-forms* $\{\omega_1, \ldots, \omega_k\}$, *independent over* $\mathcal{K}(\delta)$, *is strongly integrable if and only if there exists an index* $p \leq s(k-1)$ *such that starting from* I_0^p *defined by (3.10), the derived flag algorithm (3.9) converges to* I_∞^p *given by (3.11) with*

$$\omega_i \in \mathrm{span}_{\mathcal{K}(\delta)}\{d\varphi_1^p, \ldots, d\varphi_{\gamma_p}^p\} \quad (3.13)$$

for $i = 1, \ldots, k$.

Proof *Necessity.* If a set of one-forms $\{\omega_1, \ldots, \omega_k\}$, independent over $\mathcal{K}(\delta)$, is strongly integrable, then there exist k functions φ_i, $i = 1, \ldots, k$, such that $\mathrm{span}_{\mathcal{K}(\delta)}\{\omega_1, \ldots, \omega_k\} = \mathrm{span}_{\mathcal{K}(\delta)}\{d\varphi_1, \ldots, d\varphi_k\}$.

Thus, $\omega_i \in \mathrm{span}_{\mathcal{K}(\delta)}\{d\varphi_1, \ldots, d\varphi_k\}$ and

$$d\varphi_i \in \mathrm{span}_{\mathcal{K}}\{\omega_1, \ldots, \omega_k, \ldots, \omega_1(-p), \ldots, \omega_k(-p)\}$$

for $i = 1, \ldots, k$ and some $p \geq 0$. Clearly, $d\varphi_i \in I_\infty^p$ and Condition (3.2) is satisfied.

It remains to show that $p \leq s(k-1)$. Note that there exist infinitely many pairs $(A(\delta), \varphi)$ that satisfy $\omega = A(\delta)\mathrm{d}\varphi$. Since the degree of unimodular matrices $A(\delta)$ has a lower bound, then one can find a pair $(A(\delta), \varphi)$, where the degree of matrix $A(\delta)$ is minimal among all possible pairs. Let $A(\delta)$ be such a unimodular matrix for some functions $\varphi = (\varphi_1, \ldots, \varphi_k)^T$. Note that $A(\delta)$ and φ are not unique.

We show that the degree of $A(\delta)$ is less or equal to s. By contradiction, assume that the degree of $A(\delta)$ is larger than s, for example, $s+1$. Then for some i

$$\omega_i = a_1^i(\delta)\mathrm{d}\varphi_1 + \cdots + a_k^i(\delta)\mathrm{d}\varphi_k, \tag{3.14}$$

where $a_j^i(\delta) \in \mathcal{K}(\delta]$, $j = 1, \ldots, k$, and at least one polynomial $a_j^i(\delta)$ has degree $s+1$.

Let $a_j^i(\delta) = \sum_{l=0}^{s+1} a_{j,l}^i \delta^l$, $j = 1, \ldots, k$. From (3.14), one gets

$$\omega_i = \sum_{j=1}^{k} \sum_{\ell=0}^{s+1} a_{j,\ell}^i \mathrm{d}\varphi_j(-i), \tag{3.15}$$

where at least one coefficient $a_{j,s+1}^i \in \mathcal{K}$ is nonzero. For simplicity, assume that $a_{1,s+1}^i \neq 0$ and $a_{\gamma,s+1}^i = 0$ for $\gamma = 2, \ldots, k$. We have assumed that the maximum delay in ω_i is s, but the maximum delay in $\mathrm{d}\varphi_1(-s-1)$ is at least $s+1$.

Note that $\mathrm{d}\varphi_1, \ldots, \mathrm{d}\varphi_1(-s-1), \ldots, \mathrm{d}\varphi_k(-s-1)$ arc independent over $\mathcal{K}$. Therefore, to eliminate $\mathrm{d}\varphi_1(-s-1)$ from (3.15)

$$\mathrm{d}\varphi_1(-s-1) = \sum_{j=1}^{k} b_j(\delta)\mathrm{d}\varphi_j + \bar{\omega} \tag{3.16}$$

for some coefficients $b_j(\delta) \in \mathcal{K}(\delta]$. Let $l_j = \deg b_j(\delta) \leq s$ and the one-forms $\bar{\omega} \in \mathrm{span}_{\mathcal{K}}\{\mathrm{d}x, \mathrm{d}x^-, \ldots, \mathrm{d}x^{-s}\}$. Let $l := \min\{l_j\}$. For clarity, let $l = l_2$ and $b_2(\delta) = \delta^l$. We show that $\bar{\omega}$ can be chosen such that it is integrable. By contradiction, assume that $\bar{\omega}$ cannot be chosen integrable. Then, the coefficients of $\bar{\omega}$ must depend on larger delays than s. Since $\bar{\omega}$ is not integrable, then the coefficients of $a_{1,s+1}^i \bar{\omega}$ depend also on larger delays than s. Now, substitute $a_{1,s+1}^i \mathrm{d}\varphi_1^{-s-1}$ to (3.15). One gets that ω_i depends on $a_{1,s+1}^i \bar{\omega}$ and thus also on larger delays than s. This is a contradiction and thus $\bar{\omega}$ can be chosen integrable.

Let $\bar{\omega} = a\mathrm{d}\phi^{-l}$ for some $a, \phi \in \mathcal{K}$. Then $\mathrm{span}_{\mathcal{K}(\delta]}\{\mathrm{d}\varphi_1, \ldots, \mathrm{d}\varphi_k\} = \mathrm{span}_{\mathcal{K}(\delta]}\{\mathrm{d}\varphi_1, \mathrm{d}\phi, \mathrm{d}\varphi_3, \ldots, \mathrm{d}\varphi_k\}$ and there exists an unimodular matrix $\bar{A}(\delta)$ with smaller degree than $A(\delta)$, and functions $\bar{\varphi} = (\varphi_1, \phi, \varphi_3, \ldots, \varphi_k)^T$ such that $\omega = \bar{A}(\delta)\mathrm{d}\bar{\varphi}$, which leads to a contradiction. Thus, the degree of $A(\delta)$ must be less than or equal to s and the degree of $A^{-1}(\delta)$ is less or equal to $s(k-1)$, i.e. $p \leq s(k-1)$. The general case requires a more technical proof.

Sufficiency. Let $I_\infty^p = \text{span}_\mathcal{K}\{d\varphi\}$, where $p \le s(k-1)$. By construction, $I_\infty^p \subset \text{span}_{\mathcal{K}(\delta)}\{\omega_1, \ldots, \omega_k\}$ and by (3.13) $\omega_i \in \text{span}_{\mathcal{K}(\delta)}\{d\varphi\}$ for $i = 1, \ldots, k$. Thus, $\text{span}_{\mathcal{K}(\delta)}\{\omega_1, \ldots, \omega_k\} = \text{span}_{\mathcal{K}(\delta)}\{d\varphi\}$.

Since $I_\infty^p \subseteq I_\infty^{p+1}$ for any $p \ge 0$, one can check the condition (3.13) step by step, increasing the value of p every step. When for some $p = \bar{p}$ the condition (3.13) is satisfied, then it is satisfied for all $p > \bar{p}$.

Given the set of 1-forms $\{\omega_1, \ldots, \omega_k\}$, independent over $\mathcal{K}(\delta)$, the basis of vector space $I_\infty^{s(k-1)}$ defines the basis for the largest integrable left-submodule contained in $\text{span}_{\mathcal{K}(\delta)}\{\omega_1, \ldots, \omega_k\}$.

Lemma 3.2 *A set of one-forms $\{\omega_1, \ldots, \omega_k\}$ is weakly integrable if and only if the left closure of the left-submodule, generated by $\{\omega_1, \ldots, \omega_k\}$, is (strongly) integrable.*

Proof *Necessity.* By definitions of weak integrability and left closure, there exist functions $\varphi = (\varphi_1, \ldots, \varphi_k)^T$ such that $d\varphi = A(\delta)\bar{\omega}$, where $\bar{\omega}$ is the basis of the closure of the left-submodule, generated by $\{\omega_1, \ldots, \omega_k\}$. Choose $\{d\varphi_1, \ldots, d\varphi_k\}$ such that for $i = 1, \ldots, k$

$$d\varphi_i \ne ad\phi + \sum_{j=1; j\ne i}^{k} b_j(\delta)d\varphi_j \qquad (3.17)$$

for any $\phi \in \mathcal{K}$ and $b_j(\delta) \in \mathcal{K}(\delta)$. It remains to show that one can choose φ such that $\bar{\omega}_i \in \text{span}_{\mathcal{K}(\delta)}\{d\varphi\}$.

By contradiction, assume that one cannot choose φ such that $\bar{\omega}_i \in \text{span}_{\mathcal{K}(\delta)}\{d\varphi\}$. Then $\bar{\omega}_k \notin \text{span}_{\mathcal{K}(\delta)}\{d\varphi\}$ and also $\bar{\omega}_k^{-j} \notin \text{span}_{\mathcal{K}(\delta)}\{d\varphi_1, \ldots, d\varphi_k\}$ for $j \ge 1$ and any φ. Really, if

$$\bar{\omega}_k^{-j} = \sum_i c_i(\delta)d\varphi_i, \qquad (3.18)$$

then, since on the left-hand side of (3.18) everything is delayed at least j times, everything that is delayed less than j times on the right-hand side should cancel out. Therefore, one is able to find functions $\phi_i, \psi_i \in \mathcal{K}$, $i = 1, \ldots, k$, such that $d\varphi_i = d\phi_i + d\psi_i$ and

$$c_i(\delta)d\phi_i \in \text{span}_{\mathcal{K}(\delta)}\{dx^{-j}\} \qquad \sum_i c_i(\delta)d\psi_i = 0.$$

Now, because of (3.17), $\psi_i = 0$, $\phi_i = \varphi_i$ for $i = 1, \ldots, k$ and thus $\delta^j \bar{\omega}_k = \delta^j \sum_i \bar{c}_i(\delta)d\varphi_i^{+j}$ which yields $\bar{\omega}_k = \sum_i \bar{c}_i(\delta)d\varphi_i^{+j}$. Clearly, the one-forms $d\varphi_i^{+j}$ have to belong to $\text{span}_{\mathcal{K}(\delta)}\{\bar{\omega}\}$, because $d\varphi_i \in \text{span}_{\mathcal{K}(\delta)}\{\bar{\omega}\}$. Now, one has a contradiction and therefore $\bar{\omega}_k^{-j} \notin \text{span}_{\mathcal{K}(\delta)}\{d\varphi\}$ for $j \ge 1$. Then, by construction $\text{span}_{\mathcal{K}(\delta)}\{d\varphi_1, \ldots, d\varphi_k\} \subset \text{span}_{\mathcal{K}(\delta)}\{\omega_1, \ldots, \omega_{k-1}\}$, which is impossible. Thus, the assumption that one cannot choose φ such that $\bar{\omega}_i \in \text{span}_{\mathcal{K}(\delta)}\{d\varphi\}$ must be wrong.

Sufficiency. Sufficiency is satisfied directly by the definitions of strong and weak integrability.

Example 3.8 Consider the following one-forms:

$$\omega_1 = x_3(t-1)dx_2(t) + x_2(t)dx_3(t-1) + x_2(t-1)dx_1(t-1)$$
$$\omega_2 = x_3(t-2)dx_2(t-1) + x_2(t-1)dx_3(t-2)$$
$$+dx_1(t) + x_2(t-2)dx_1(t-2). \tag{3.19}$$

One gets for $s(k-1) = 2$:

$$I_\infty^2 = \text{span}_{\mathcal{K}}\{dx_1(t), dx_1(t-1), d(x_2(t)x_3(t-1))\}.$$

When one eliminates the basis elements, which are dependent over $\mathcal{K}(\delta]$, one gets that the rank of $\text{span}_{\mathcal{K}(\delta]}\{dx_1(t), dx_1(t-1), d(x_2(t)x_3(t-1))\}$ is 2. To check the condition (3.13), one has to check whether there exists a matrix $A(\delta)$ such that $\omega = A(\delta)d\varphi$, where $\omega = (\omega_1, \omega_2)^T, \varphi = (\varphi_1, \varphi_2)^T, \varphi_1 = x_2(t)x_3(t-1)$, and $\varphi_2 = x_1(t)$. In fact, $\omega = A(\delta)d\varphi$, where the unimodular matrix $A(\delta)$ is

$$A(\delta) = \begin{bmatrix} 1 & x_2(t-1)\delta \\ \delta & 1 + x_2(t-2)\delta^2 \end{bmatrix}.$$

Thus, the one-forms (3.19) are strongly integrable.

Example 3.9 Consider the following one-forms:

$$\omega_1 = dx_2(t)$$
$$\omega_2 = x_4(t-1)dx_1(t) + x_2(t)dx_2(t-1) + x_1(t)dx_4(t-1)$$
$$\omega_3 = x_3(t)x_4(t)dx_2(t) + x_2(t)x_4(t)dx_3(t) \tag{3.20}$$
$$+x_3(t-1)dx_2(t-1) + x_2(t-1)dx_3(t-1).$$

For $s(k-1) = 2$:

$$I_\infty^2 = \text{span}_{\mathcal{K}}\{dx_2(t), d(x_4(t-1)x_1(t)), dx_2(t-1), dx_2(t-2), d(x_4(t-2)x_1(t-1))\}.$$

Now, $\omega_1 \in I_\infty^2$ and $\omega_2 \in I_\infty^2$, but $\omega_3 \notin I_\infty^2$. Thus, the one-forms (3.20) are not strongly integrable, and $\text{span}_{\mathcal{K}(\delta]}\{dx_2(t), d(x_4(t-1)x_1(t))\}$ is the largest integrable left-submodule, contained in $\mathcal{A} = \text{span}_{\mathcal{K}(\delta]}\{\omega_1, \omega_2, \omega_3\}$.

Now, one can check if the one-forms (3.20) are weakly integrable. For that, one has to compute the left closure of $\mathcal{A}$ and check if it is strongly integrable. In practice, the left closure of a left-submodule $\mathcal{A}$ can be computed as the left-kernel of its right-kernel Δ. Thus, the right-kernel of $\mathcal{A}$ is $\Delta = \text{span}_{\mathcal{K}(\delta]}\{q(\delta)\}$, where $q(\delta) = (x_1(t)\delta, 0, 0, -x_4(t))^T$. The left-kernel of Δ is

$$cl_{\mathcal{K}(\delta]}(\mathcal{A}) = \text{span}_{\mathcal{K}(\delta]}\{dx_2(t), dx_3(t), d(x_4(t-1)x_1(t))\}.$$

Therefore, the one-forms (3.20) are weakly integrable.

We finally end this section by highlighting that the integrability of right-sub modules and one-forms are connected through the following corollary, which follows from Lemma 3.2 and by noting also that given a left-submodule $\mathcal{A}$, its right-kernel, and the right-kernel of its closure coincide and analogously given a right-submodule Δ, its left-kernel, and the left-kernel of its closure coincide.

Corollary 3.2 *Weak integrability of a set of one-forms is equivalent to the integrability of its right-kernel.*

To show more explicitly how the integrability of right-submodules and weak integrability of one-forms are related, consider Algorithm (3.9) initialized with (3.12). The left-kernel of Δ_i, defined above, is equal to I_∞^i, where I_∞^i is computed with respect to the closure of a given submodule.

The next example shows the importance of working on $\mathcal{K}^*$ and $\mathcal{K}^*(\delta]$. In fact, while the one-forms considered are causal, the right-kernel is not thus requiring to consider the polynomial Lie Bracket on $\mathcal{K}^*(\delta]$ as defined in Definition 2.4 to check the integrability.

Example 3.10 Consider the one-forms

$$
\begin{aligned}
\omega_1 &= x_1(t-1)\mathrm{d}x_1(t) + x_1(t)\mathrm{d}x_1(t-1) \\
&\quad -x_3(t)\mathrm{d}x_2(t-1) + \mathrm{d}x_3(t-1) \\
\omega_2 &= \mathrm{d}x_2(t) + x_3(t)\mathrm{d}x_2(t-1).
\end{aligned}
\tag{3.21}
$$

The one-forms $\omega = (\omega_1, \omega_2)^T$ can be written as

$$
\omega = \begin{pmatrix} x_1(t-1) + x_1(t)\delta & -x_3(t)\delta & \delta \\ 0 & 1 + x_3(t)\delta & 0 \end{pmatrix} \mathrm{d}x(t).
$$

The right-kernel of the left-submodule $\mathrm{span}_{\mathcal{K}(\delta]}\{\omega_1, \omega_2\}$ is not causal (*i.e.* one needs forward-shifts of variables $x(t)$ to represent it), and is given by

$$
\Delta = \mathrm{span}_{\mathcal{K}^*(\delta]} \left\{ \begin{pmatrix} x_1(-1)\delta \\ 0 \\ -x_1^2(0) - x_1(1)x_1(-1)\delta \end{pmatrix} \right\} = \mathrm{span}_{\mathcal{K}^*(\delta]}\{r_1(\mathbf{x}, \delta)\}.
$$

To check the involutivity of Δ, we have to consider

$$
R_1(\mathbf{x}, \epsilon) = \begin{pmatrix} x_1(-1)\epsilon(-1) \\ 0 \\ -x_1^2(0)\epsilon(0) - x_1(1)x_1(-1)\epsilon(-1) \end{pmatrix}
$$

from which since $s_1 = 1$, we compute the Polynomial Lie Bracket

$$[R_1(\mathbf{x}, \epsilon), r_1(\mathbf{x}, \delta)] = \dot{r}_1(\mathbf{x}, \delta)|_{\dot{\mathbf{x}}_{[0]} = R_1(\mathbf{x}, \epsilon)\delta} - \sum_{k=0}^{2} \frac{\partial R_1(\mathbf{x}, \epsilon)}{\partial x(1-k)} \delta^k r_1(\mathbf{x}(s_1), \delta)$$

$$= \begin{pmatrix} x_1(-2)\delta^2 \\ 0 \\ -x_1(0)x_1(-1)\delta - x_1(1)x_1(-2)\delta^2 \end{pmatrix} (\epsilon(0) - \delta\epsilon(2)).$$

Since

$$\begin{pmatrix} x_1(-2)\delta^2 \\ 0 \\ -x_1(0)x_1(-1)\delta - x_1(1)x_1(-2)\delta^2 \end{pmatrix} = r_1(\mathbf{x}, \delta) \frac{x_1(-1)}{x_1} \delta,$$

Δ is involutive which shows that the closure of ω is strongly integrable. In fact, one gets that the left-annihilator of Δ is generated by

$$\Delta^{\perp} = \mathrm{span}_{\mathcal{K}^*(\delta]}\{d(x_1(t)x_1(t-1) + x_3(t-1)), dx_2(t)\} = \mathrm{span}_{\mathcal{K}^*(\delta]}\{d\lambda_1, d\lambda_2\}.$$

On the other hand

$$\begin{pmatrix} \omega_1 \\ \omega_2 \end{pmatrix} = \begin{pmatrix} 1 & -x_3\delta \\ 0 & 1+x_3\delta \end{pmatrix} \begin{pmatrix} d\lambda_1 \\ d\lambda_2 \end{pmatrix},$$

which shows that $\mathrm{span}_{\mathcal{K}^*(\delta]}\{\omega_1, \omega_2\} \subset \mathrm{span}_{\mathcal{K}^*(\delta]}\{d\lambda_1, d\lambda_2\}$, that is, weak integrability.

The problem can be addressed by working directly on one-forms. In this case, starting from ω_1 and ω_2, we first consider the left closure which is given by

$$\Omega_c = \mathrm{span}_{\mathcal{K}^*(\delta]}\{\omega_1, dx_2\}$$

and then we can apply the derived flag algorithm, thus getting

$$\Omega_c = \mathrm{span}_{\mathcal{K}^*(\delta]}\{d(x_1(t)x_1(t-1) + x_3(t-1)), dx_2(t)\} = \mathrm{span}_{\mathcal{K}^*(\delta]}\{d\lambda_1, d\lambda_2\}$$

as expected.

3.4 Problems

1. Consider the one-form $\omega = dx(t) + x(t)dx(t-1)$. Check whether it is weakly integrable and/or strongly integrable.

2. Check if the right-submodule

$$\Delta(\delta] = \mathrm{span}_{\mathcal{K}^*(\delta]} \left\{ \begin{pmatrix} x_1(0)x_1(1)\delta \\ -x_1(0)x_1(2) - x_1^2(0)\delta \end{pmatrix} \right\}$$

is 0-integrable.

3. Check if the right-submodule

$$\Delta(\delta] = \text{span}_{\mathcal{K}^*(\delta]} \left\{ \begin{pmatrix} x_1(2)x_1(3)\delta \\ -x_1(2)x_1(4) - x_1^2(2)\delta \end{pmatrix} \right\}$$

is p-integrable for some $p \geq 0$.

4. Prove the following lemma:

Lemma 3.3 *Consider the distribution Δ_i defined by (3.8), and let $\rho_i = dim(\Delta_1)$ with $\rho_{-1} = ns$. Then*

(i) *If $d\lambda(\mathbf{x})$ is such that $\text{span}\{d\lambda(\mathbf{x})\} = \bar{\Delta}_{i-1}^\perp$, then $\text{span}\{d\lambda(\mathbf{x}), d\lambda(\mathbf{x}(-1))\} \subset \bar{\Delta}_i^\perp$.*

(ii) *A canonical basis for $\bar{\Delta}_i^\perp$ is defined for $i \geq 0$ as follows:*
   Pick $d\lambda_0(\mathbf{x}_{[0]})$ such that $\text{span}\{d\lambda_0(\mathbf{x}_{[0]})\} = \bar{\Delta}_0^\perp$, with rank $d(\lambda_0) = \mu_0 = \rho_0 - \rho_{-1}$.
   At step $\ell \leq i$ pick $d\lambda_\ell(\mathbf{x}_{[\ell]})$ such that $\text{span}\{d\lambda_k(\mathbf{x}_{[k]}(-j)), k \in [0, \ell], j \in [0, \ell - k]\} = \bar{\Delta}_\ell^\perp$, and $d\lambda_\ell(\mathbf{x}_{[\ell]}) \notin \bar{\Delta}_{\ell-1}^\perp$, with rank $d(\lambda_i) = \mu_i = \rho_i - 2\rho_{i-1} + \rho_{i-2}$.

Chapter 4
Accessibility of Nonlinear Time-Delay Systems

In this chapter, the accessibility properties of a nonlinear time-delay system affected by constant commensurate delays are fully characterized in terms of absence of non-constant autonomous functions, that is, functions which are non-constant and whose derivatives of any order are never affected by the control. Using an algebraic terminology, the accessibility property is characterized by the accessibility module $\mathcal{R}_n$ introduced in Márquez-Martínez (1999) and defined by (4.8) to be torsion free over the ring $\mathcal{K}(\delta]$. This has been worked out in Fliess and Mounier (1998) for the special case of linear time-delay systems. While controllability for linear time-delay systems was first addressed in Buckalo (1968), a complete characterization in the linear case was instead given in Sename (1995).

In order to understand the peculiarities of time-delay systems, let us first analyze the following example:

Example 4.1 Consider the delay-free second-order nonlinear system in chained form

$$\dot{x}_1(t) = x_2(t)u(t)$$
$$\dot{x}_2(t) = u(t).$$

It is well known that such a system is not locally accessible. The accessibility distribution associated to it is $\mathcal{R}_2 = \text{span}\left\{ \begin{pmatrix} x_2 \\ 1 \end{pmatrix} \right\}$ which has dimension 1 for any x. As a matter of fact, the function $\varphi = x_1(t) - \frac{1}{2}x_2^2(t)$ is an autonomous function for the given system and it is computed starting from $\mathcal{R}_2$.

Surprisingly, a delay on x_2 renders the system locally accessible. In fact, as shown in Califano et al. (2013), the nonlinear system

© The Author(s), under exclusive license to Springer Nature Switzerland AG 2021
C. Califano and C. H. Moog, *Nonlinear Time-Delay Systems*,
SpringerBriefs in Control, Automation and Robotics,
https://doi.org/10.1007/978-3-030-72026-1_4

$$\dot{x}_1(t) = x_2(t - 1)u(t)$$
$$\dot{x}_2(t) = u(t)$$

(4.1)

is locally accessible (there is no way to compute an autonomous function for such a system). This is discussed in Example 4.3. Using the results obtained in this chapter it is shown that the rank of the accessibility submodule associated to a given delay system determines the dimension of its accessible subsystem and consequently the dimension of its non-accessible part.

It should be noted that a different slight modification in the dynamics (4.1) yields again a non-accessible system.

Example 4.2   Consider, for instance,

$$\dot{x}_1(t) = x_2(t - 1)u(t - 1)$$
$$\dot{x}_2(t) = u(t).$$

(4.2)

Now, an autonomous function is computed as $\varphi = x_1(t) - \frac{1}{2}x_2^2(t - 1)$ since $\dot{\varphi} = 0$.
   As a matter of fact, in the coordinates

$$\begin{pmatrix} z_1(t) \\ z_2(t) \end{pmatrix} = \begin{pmatrix} x_1(t) - \frac{1}{2}x_2^2(t - 1) \\ x_2(t) \end{pmatrix}$$

(4.3)

the system reads

$$\dot{z}_1(t) = 0$$
$$\dot{z}_2(t) = u(t)$$

which by the way is also delay free.

   This means that the points reachable at time t are linked to the point reached at time $t - 1$ through the relation $x_1(t) - \frac{1}{2}x_2^2(t - 1) = constant$: the trajectory of the state is thus constrained by the initialization. However, it is still possible to reach a given fixed point in $I\!R^2$ at a different time $\bar{t}$, even if the previous link is present between the point reached at time $t - 1$ and that one reached at time t. In fact, assume to start the system with $x_1(t) = x_{10}$, $x_2(t) = x_{20}$, $u(t) = 0$ for $t \in [-1, 0)$, and let $x_f = (x_{1f}, x_{2f})^T$ be the final point to be reached.

• On the interval $[0, 1)$, by setting $u(t) = u_0$, one gets that

$$x_1(t) = x_{10}$$
$$x_2(t) = u_0 t + x_{20}.$$

If $x_{1f} \neq x_{10}$ it is immediately clear that any arbitrary final point x_f cannot be reached for $t \in [0, 1)$. The set of reachable points from x_0, within some time $t \in [0, 1)$, is not open in $I\!R^2$.

- Now let the control switch to $u(t) = u_1$ on the interval $t \in [1, 2)$. The dynamics on such interval becomes

$$\dot{x}_1(t) = x_2(t-1)u_0 = ((t-1)u_0 + x_{20})u_0$$
$$\dot{x}_2(t) = u_1.$$

The solution is obtained as

$$x_1(t) = \frac{1}{2}(t-1)^2 u_0^2 + (t-1)x_{20}u_0 + x_{10}$$
$$x_2(t) = (t-1)u_1 + x_2(1) = (t-1)u_1 + u_0 + x_{20}$$

so that once the time $\bar{t} \in [1, 2)$ at which one wants to reach the desired state x_f is fixed, one gets that

$$x_{1f}(\bar{t}) = \frac{1}{2}(\bar{t}-1)^2 u_0^2 + (\bar{t}-1)x_{20}u_0 + x_{10}$$
$$x_{2f}(\bar{t}) = (\bar{t}-1)u_1 + u_0 + x_{20}.$$

Accordingly, with the control sequence

$$u_0 = \frac{-x_{20} \pm \sqrt{x_{20}^2 - 2(x_{10} - x_{1f})}}{(\bar{t}-1)}$$

$$u_1 = \frac{x_{2f} - x_{20} - u_0}{(\bar{t}-1)}$$

one can reach any final point x_f from the initial point x_0 as long as $x_{20}^2 - 2(x_{10} - x_{1f}) \geq 0$.

Of course using non-constant controls may allow to reach a greater region of the plane. However, already this simple control shows, on one hand, that the system is accessible in a weaker sense (the set of reachable points is open), and, on the other hand, that while it is possible to get through the desired final point x_f, it is not, however, possible to stay in x_f forever.

This example shows the necessity of introducing a weaker notion of accessibility that we call *t-accessibility*. We end this introductory paragraph by giving the formal definition of accessible and t-accessible system which will be used later on in the chapter.

Consider the dynamics

$$\dot{\mathbf{x}}_{[0]} = F(\mathbf{x}_{[s]}) + \sum_{i=0}^{l} \sum_{j=1}^{m} G_{ji}(\mathbf{x}_{[s]}) u_{j,[0]}(-i). \tag{4.4}$$

Definition 4.1 The state $x_f \in I\!R^n$ is said to be reachable from the initial condition $\varphi(t)$, $t \in [-s\tau, 0)$, if there exist a time t_f and a Lebesgue measurable input $u(t)$ defined for $t \in [0, t_f]$ such that the solution $x(t_f, \varphi)$ of (4.4) equals x_f.

Definition 4.2 System (4.4) is said to be t-accessible from the initial condition $\varphi(t)$, $t \in [-s\tau, 0)$, if the adherence of the set of its reachable states has a nonempty interior.

There may exist some *singular* initial conditions where the system is not t-accessible. Thus, system (4.4) is just said to be t-accessible if it is t-accessible from almost any initial condition.

Definition 4.3 System (4.4) is fully accessible if there does not exist any autonomous function for the system, that is, a non-constant function $\lambda(\mathbf{x})$ whose time derivative of any order along the dynamics of the system is never affected by the control.

4.1   The Accessibility Submodules in the Delay Context

To tackle the accessibility problem for a given dynamics of the form (4.4), we are essentially looking for a bicausal change of coordinates $\mathbf{z}_{[0]} = \phi(\mathbf{x})$ such that in the new coordinates the system is split into two subsystems

$$\dot{z}_{1,[0]} = \tilde{F}_1(z_{1,[\bar{s}]})$$

$$\dot{z}_{2,[0]} = \tilde{F}_2(\mathbf{z}) + \sum_{i=0}^{\bar{l}} \sum_{j=1}^{m} \tilde{G}_{2,ji}(\mathbf{z}) u_{j,[0]}(-i)$$

with the subsystem S_2 completely accessible, that is, satisfying definition 4.3.

In order to address this problem, we need to refer to the differential representation of the given dynamics (4.4), which was introduced in Sect. 2.3. Thus, by applying the differential operator d to both sides of (4.4), its differential form representation is derived as

$$d\dot{\mathbf{x}}_{[0]} = f(\mathbf{x}_{[s]}, \mathbf{u}_{[0]}, \delta) d\mathbf{x}_{[0]} + \mathbf{g}_1(\mathbf{x}_{[s]}, \delta) d\mathbf{u}_{[0]}, \tag{4.5}$$

where we recall that

$$f(\mathbf{x}, \mathbf{u}, \delta) = \sum_{i=0}^{s} \left(\frac{\partial F(\mathbf{x})}{\partial \mathbf{x}_{[0]}(-i)} + \sum_{j=1}^{m} \sum_{k=0}^{l} u_{j,[0]}(-k) \frac{\partial G_{jk}(\mathbf{x})}{\partial \mathbf{x}_{[0]}(-i)} \right) \delta^i, \tag{4.6}$$

$$\mathbf{g}_1(\mathbf{x}, \delta) = (g_{11}, \dots, g_{1m}), \quad g_{1i} = \sum_{k=0}^{l} G_{ik}(\mathbf{x}_{[s]}) \delta^k, \ i \in [1, m]. \tag{4.7}$$

We will assume, without loss of generality, that $\mathrm{rank}_{\mathcal{K}(\delta]}(\mathbf{g}_1(\mathbf{x}, \delta)) = m$ (the number of inputs), that is, each input acts independently on the system.

The first step consists in finding out the maximum number of independent autonomous functions for the given system and then in showing that these functions can be used to define a bicausal change of coordinates. To this end, we need to characterize the notion of relative degree, since as it will be stated later on, and in accordance with the delay-free case, autonomous functions have actually infinite relative degree. Starting from the g_{1i}'s defined by (4.7), we thus consider the accessibility submodule generators, introduced in Márquez-Martínez (1999, 2000), defined (up to the sign) as

$$g_{i+1,j}(\mathbf{x}, \mathbf{u}^{[i-1]}, \delta) = \dot{g}_{i,j}(\mathbf{x}, \mathbf{u}^{[i-2]}, \delta) - f(\mathbf{x}, \mathbf{u}, \delta)g_{i,j}(\mathbf{x}, \mathbf{u}^{[i-2]}, \delta).$$

The accessibility submodules $\mathcal{R}_i$ of Σ again introduced in Márquez-Martínez (1999) are then defined as

$$\mathcal{R}_i(\mathbf{x}, \mathbf{u}^{[i-2]}, \delta) = \mathrm{span}_{\mathcal{K}(\delta]}\{\mathbf{g}_1(\mathbf{x}, \delta) \ldots \mathbf{g}_i(\mathbf{x}, \mathbf{u}^{[i-2]}, \delta)\}. \tag{4.8}$$

Let us now recall that a function $\lambda(\mathbf{x}_{[\bar{s}]})$ has finite relative degree k if $\forall l \in [1, m]$, and $\forall i \in [1, k-1]$

$$L_{g_{il}^j(\mathbf{x}, \mathbf{u}^{[i-2]})}\lambda(\mathbf{x}_{[\bar{s}]}) = 0, \ \forall j \in [0, \bar{s} + \beta_i], \ \forall \mathbf{u}^{[i-2]}, \tag{4.9}$$

and there exists an index $l \in [1, m]$ such that

$$L_{g_{kl}^j(\mathbf{x}, \mathbf{u}^{[k-2]})}\lambda(\mathbf{x}_{[\bar{s}]}) \neq 0 \text{ for some } j \in [0, \bar{s} + \beta_k]. \tag{4.10}$$

It immediately follows that a function $\lambda(\mathbf{x})$ has relative degree $k > 1$ if and only if

$$d\lambda(\mathbf{x}) \perp \mathcal{R}_{k-1}(\mathbf{x}, \mathbf{u}^{[k-3]}, \delta)$$
$$d\lambda(\mathbf{x})g_{k,\ell}(\mathbf{x}, \mathbf{u}^{[k-2]}, \delta) \neq 0 \text{ for some } \ell \in [1, m]. \tag{4.11}$$

The following result, which is an immediate consequence of the expression of the $g_{il}(\mathbf{x}, \mathbf{u}, \delta)$'s given later on by (4.22), gives conditions which are independent of the control u, for a function to have relative degree k.

Proposition 4.1 *A function $\lambda(\mathbf{x})$ has relative degree $k > 1$ if and only if $\forall l \in [1, m]$,*

$$d\lambda(\mathbf{x})g_{il}(\mathbf{x}, 0, \delta) = 0, \forall i \leq k - 1, \tag{4.12}$$

and for some $l \in [1, m]$,

$$d\lambda(\mathbf{x})g_{kl}(\mathbf{x}, 0, \delta) \neq 0. \tag{4.13}$$

A straightforward consequence of the definition of relative degree and accessibility submodules is that a non-constant function has infinite relative degree if and only if its relative degree is greater than n, which also allows to characterize autonomous functions. We have, in fact, the following results which allow to derive an accessibility criterion.

Lemma 4.1 *Given the dynamics (4.4), the relative degree of a non-constant function* $\lambda(\mathbf{x}_{[\bar{s}]}) \in \mathcal{K}$ *is greater than n if and only if it is infinite.*

Theorem 4.1 *The dynamics (4.4) is locally accessible if and only if the following equivalent statements hold true:*

- $\mathcal{R}_n(\mathbf{x}, \mathbf{u}^{[n-2]}, \delta)$ *is torsion free over* $\mathcal{K}(\delta]$,
- $\text{rank}_{\mathcal{K}(\delta]} \mathcal{R}_n(\mathbf{x}, \mathbf{u}^{[n-2]}, \delta) = n$ *for some* $\mathbf{u}^{[n-2]}$, *and*
- $\text{rank}\, \mathcal{R}_n(\mathbf{x}, 0, \delta) = n$.

Proof If $\mathcal{R}_n(\mathbf{x}, \mathbf{u}^{[n-2]}, \delta)$ is torsion free over $\mathcal{K}(\delta]$, then there is no nonzero element which annihilates $\mathcal{R}_n(\mathbf{x}, \mathbf{u}^{[n-2]}, \delta)$, that is, $\text{rank}_{\mathcal{K}(\delta]} \mathcal{R}_n(\mathbf{x}, \mathbf{u}^{[n-2]}, \delta) = n$. Consequently, there cannot exist any function with infinite relative degree, $\text{rank}\, \bar{\mathcal{R}}(\mathbf{x}, 0, \delta) = n$ and the given system is accessible. As for the converse, assume that $\mathcal{R}_n(\mathbf{x}, \mathbf{u}^{[n-2]}, \delta)$ is not torsion free over $\mathcal{K}(\delta]$. Then $\text{rank}_{\mathcal{K}(\delta]} \mathcal{R}_n(\mathbf{x}, \mathbf{u}^{[n-2]}, \delta) = k < n$ for all possible choices of $\mathbf{u}^{[k-2]}$. Accordingly (see Proposition 4.4), $\bar{\mathcal{R}}_n(\mathbf{x}, 0, \delta]$ the involutive closure of $\mathcal{R}_n(\mathbf{x}, 0, \delta]$ has rank k, so that there exist $n - k$ exact differentials in the left-annihilator, independent over $\mathcal{K}(\delta]$, which would contradict the assumption that the system is fully accessible since for Proposition 4.1 the corresponding functions have infinite relative degree.

Example 4.3 (*The Chained Form Model*) Consider again the two-dimensional system (1.8):

$$\dot{\mathbf{x}}_{[0]} = g(\mathbf{x}_{[1]})u_{[0]} = \begin{pmatrix} x_{2,[0]}(-1) \\ 1 \end{pmatrix} u_{[0]}.$$

As already discussed in Sect. 2.5 and showed in Fig. 1.2, when the delay $\tau \neq 0$, the system becomes fully accessible, as opposite to the delay-free case (Bloch 2003; Murray and Sastry 1993; Sørdalen 1993). In fact, through standard computations, one has that

$$g_1(\mathbf{x}, \delta) = \begin{pmatrix} x_{2,[0]}(-1) \\ 1 \end{pmatrix}, \quad g_2(\mathbf{x}, \mathbf{u}, \delta) = \begin{pmatrix} u_{[0]}(-1) - u_{[0]}\delta \\ 0 \end{pmatrix},$$

which shows that the accessibility matrix

$$\mathcal{R}_2(\mathbf{x}, \mathbf{u}, \delta) = \begin{pmatrix} x_{2,[0]}(-1) & u_{[0]}(-1) - u_{[0]}\delta \\ 1 & 0 \end{pmatrix}$$

has full rank whenever the control is different from 0, which, according to Theorem 4.1, ensures the full accessibility of the given system. An extensive discussion on this topic can be found in Califano et al. (2013), Li et al. (2011).

We end this section by noting that the accessibility property of a system can also be given working on one-forms. In fact, starting from the definition of infinite relative degree of a one-form, it turns out that a system is completely accessible if and only if the set of integrable one-forms with infinite relative degree is empty. To get the result, we first have to define the relative degree of a one-form which is given hereafter.

Definition 4.4 A one-form $\omega \in \text{span}_{\mathcal{K}(\delta]}\{dx(t)\}$ has relative degree r, if r is the smallest integer such that $\omega^{(r)} \notin \text{span}_{\mathcal{K}_u(\delta]}\{dx(t)\}$. A function $\varphi \in \mathcal{K}$ is said to have relative degree r if the one-form $d\varphi$ has relative degree r.

Denoting now by $\mathcal{M} = \text{span}_{\mathcal{K}(\delta]}\{dx(t), du^{(k)}(t); k \geq 0\}$ let us now consider the sequence of left-submodule $\mathcal{H}_1 \supset \mathcal{H}_2 \supset \ldots$ of $\mathcal{M}$ as follows:

$$\mathcal{H}_1 = \text{span}_{\mathcal{K}(\delta]}\{dx(t)\}$$
$$\mathcal{H}_i = \text{span}_{\mathcal{K}(\delta]}\{\omega \in \mathcal{H}_{i-1} \mid \dot{\omega} \in \mathcal{H}_{i-1}\}. \tag{4.14}$$

Since $\mathcal{H}_1$ has finite rank and all the left submodule $\mathcal{H}_i$ are closed it was shown in Xia et al. (2002) that the sequence (4.14) converges. Let now $\mathcal{H}_\infty$ be the limit of sequence (4.14) and let $\hat{\mathcal{H}}_i$ denote the largest integrable left submodule contained in $\mathcal{H}_i$. A left submodule $\mathcal{H}_i$ contains all the one-forms with relative degree equal to or greater than i. Thus, $\mathcal{H}_\infty$ contains all the one-forms which have infinite relative degree. As a consequence, the accessibility of system (4.4) can be characterized in the following way.

Theorem 4.2 *System (4.4) is accessible if and only if $\hat{\mathcal{H}}_\infty = \emptyset$.*

4.2 A Canonical Decomposition with Respect to Accessibility

Theorem 4.1 gives a criterion to test the accessibility of a given system. If $rank_{\mathcal{K}(\delta]} \mathcal{R}_n = k < n$ the system is not accessible and there exist $n - k$ independent functions $\varphi_1(\mathbf{x}), \ldots, \varphi_{n-k}(\mathbf{x})$ which are characterized by an infinite relative degree.

We are thus interested in characterizing the non-accessible part of the system, that is, defining a bicausal change of coordinates which decomposes in the new coordinates the given system into two parts, one of which represents the non-accessible subsystem.

The basic idea is that the autonomous functions linked to the system and that can be computed through $\mathcal{R}_n(\mathbf{x}, 0, \delta)$ represent the starting point in order to define the correct change of coordinates. The main issues which are solved in the next two results are first to show how to compute a basis over $\mathcal{K}(\delta]$ of all the autonomous functions, and secondly to show that such a basis is closed and can be used to define a bicausal change of coordinates.

Let us then consider $\mathcal{R}_n(\mathbf{x}, 0, \delta) = \text{span}_{\mathcal{K}(\delta)}\{\mathbf{g}_1(\mathbf{x}, \delta), \ldots, \mathbf{g}_n(\mathbf{x}, 0, \delta)\}$ and, since the elements of the submodule are by construction causal, consider for $i \geq 0$, the sequence of distributions $\mathcal{G}_i \subset \text{span}\left\{\frac{\partial}{\partial \mathbf{x}_{[0]}}, \ldots, \frac{\partial}{\partial \mathbf{x}_{[0]}(-i-s)}\right\}$ defined as

$$
\mathcal{G}_i = \text{span} \begin{pmatrix} \mathbf{g}^0(\mathbf{x}_{[s]}) \cdots & \mathbf{g}^\ell(\mathbf{x}_{[s]}) & 0 & & 0 \\ 0 & \ddots & & \ddots & \\ \vdots & 0 & \mathbf{g}^0(\mathbf{x}_{[s]}(-i)) \cdots & \mathbf{g}^\ell(\mathbf{x}_{[s]}(-i)) & 0 \\ 0 & \cdots & 0 & & I_{ns} \end{pmatrix}, \tag{4.15}
$$

where ℓ represents the maximum degree in δ and s the maximum delay in x which are present in the $g_{i,j}$'s. $\mathcal{G}_i$ is a distribution in $I\!\!R^{n(s+i+1)}$ as well as its involutive closure $\bar{\mathcal{G}}_i$. Let $\rho_i = \text{rank}(\bar{\mathcal{G}}_i)$, with $\rho_{-1} = ns$. The following result can be stated.

Proposition 4.2 *Assume that the system Σ, given by (4.4), is not accessible, i.e. rank $\mathcal{R}_n(\mathbf{x}, \mathbf{u}, \delta) = k < n$, then the following facts hold true:*

(i) *The system Σ possesses $n - k$ independent (over $\mathcal{K}(\delta)$) autonomous exact differentials.*

(ii) *A canonical basis for $\bar{\mathcal{G}}_i^\perp$ is defined for $i \geq 0$ as follows.*
Let $d\lambda_0(\mathbf{x}_{[0]})$ be such that $\text{span}\{d\lambda_0(\mathbf{x}_{[0]})\} = \bar{\mathcal{G}}_0^\perp$, with rank $(d\lambda_0) = \mu_0 = \rho_0 - \rho_{-1}$.
Let $d\lambda_1(\mathbf{x}_{[1]}) \notin \bar{\mathcal{G}}_0^\perp$, with rank $(d\lambda_1) = \mu_1 = \rho_1 - 2\rho_0 + \rho_{-1}$, be such that

$$\text{span}\{d\lambda_0(\mathbf{x}_{[0]}), d\lambda_0(\mathbf{x}_{[0]}(-1)), d\lambda_1(\mathbf{x}_{[1]})\} = \bar{\mathcal{G}}_1^\perp.$$

More generally, let $d\lambda_i(\mathbf{x}_{[i]}) \notin \bar{\mathcal{G}}_{i-1}^\perp$, with rank $(d\lambda_i) = \mu_i = \rho_i - 2\rho_{i-1} + \rho_{i-2}$ be such that $\text{span}\{d\lambda_\mu(\mathbf{x}_{[\mu]}(-j)), \mu \in [0, i], j \in [0, i - \mu]\} = \bar{\mathcal{G}}_i^\perp$.

(iii) *Let $\bar{\ell}$ represent the maximum degree in δ and $\bar{s}$ the maximum delay in x in $\mathcal{R}_n(\mathbf{x}, \mathbf{u}, \delta)$. Then there exists $\gamma \leq \bar{s} + k\bar{\ell}$ such that any other autonomous function $\lambda(\mathbf{x})$ satisfies*

$$d\lambda(\mathbf{x}) \in \text{span}_{\mathcal{K}(\delta)}\{d\lambda_0(\mathbf{x}), \ldots, d\lambda_\gamma(\mathbf{x})\},$$

that is, $\bar{\mathcal{G}}_\gamma$ characterizes completely all the independent autonomous functions of Σ.

Proof (i) is a direct consequence of Proposition 4.4. (ii) is a direct consequence of Lemma 3.3 in the appendix, where $\Delta_i = \mathcal{G}_i$ is causal by assumption, thus ensuring that the left-annihilator is causal also. Finally, (iii) is a direct consequence of Lemma 1.1.

Theorem 4.3 *Consider the continuous-time system (4.4). Let γ be the smallest index such that any autonomous function $\lambda(\mathbf{x})$ associated to the given system satisfies*

$$d\lambda(\mathbf{x}) \in \text{span}_{\mathcal{K}(\delta)}\{d\lambda_0(\mathbf{x}_{[0]}), \ldots, d\lambda_\gamma(\mathbf{x}_{[\gamma]})\},$$

where

$$\mathrm{span}\{d\lambda_0(\mathbf{x})\} = \bar{\mathcal{G}}_0^\perp$$
$$\mathrm{span}\{d\lambda_0, d\lambda_0(\mathbf{x}(-1)), d\lambda_1(\mathbf{x})\} = \bar{\mathcal{G}}_1^\perp, \quad d\lambda_1(\mathbf{x}_{[1]}) \notin \bar{\mathcal{G}}_0^\perp$$
$$\vdots$$
$$\mathrm{span}\{d\lambda_i(\mathbf{x}(-j)), i \in [0, \gamma], j \in [0, \gamma - i]\} = \bar{\mathcal{G}}_\gamma^\perp, \quad d\lambda_\gamma(\mathbf{x}_{[\gamma]}) \notin \bar{\mathcal{G}}_{\gamma-1}^\perp$$

then

(1.) $\exists\, d\lambda_{\gamma+1}(\mathbf{x})$ *such that*

$$dz_{[0]} = \begin{pmatrix} dz_{1,[0]} \\ \vdots \\ dz_{\gamma+1,[0]} \\ dz_{\gamma+2,[0]} \end{pmatrix} = \begin{pmatrix} d\lambda_0(\mathbf{x}_{[0]}) \\ \vdots \\ d\lambda_\gamma(\mathbf{x}_{[\gamma]}) \\ d\lambda_{\gamma+1}(\mathbf{x}) \end{pmatrix} = T(\mathbf{x}, \delta)d\mathbf{x}_{[0]}$$

defines a bicausal change of coordinates.

(2.) In the above-defined coordinates $\mathbf{z}_{[0]} = \phi(\mathbf{x})$ *such that* $d\mathbf{z}_{[0]} = T(\mathbf{x}, \delta)d\mathbf{x}_{[0]}$ *the system reads*

$$\dot{z}_{1,[0]} = f_1(z_{1,[\bar{s}]}, \ldots, z_{\gamma+1,[\bar{s}]})$$
$$\vdots$$
$$\dot{z}_{\gamma+1,[0]} = f_{\gamma+1}(z_{1,[\bar{s}]}, \ldots, z_{\gamma+1,[\bar{s}]}) \tag{4.16}$$
$$\dot{z}_{\gamma+2,[0]} = f_{\gamma+2}(\mathbf{z}) + \sum_{i=0}^{\bar{s}} \sum_{j=1}^{m} \tilde{G}_{ji}(\mathbf{z})u_{j,[0]}(-i).$$

Moreover, the dynamics associated to $(z_1, \ldots z_{\gamma+1})^T$ *represents the largest non-accessible dynamics.*

Proof By construction $\mathrm{span}_{\mathcal{K}(\delta]}\{d\lambda_0(\mathbf{x}), \ldots, d\lambda_\gamma(\mathbf{x})\}$ is closed and its right-annihilator is causal so that, according to Proposition 4.2, it is possible to compute $\lambda_{\gamma+1}(\mathbf{x})$ such that

$$\mathbf{z}_{[0]} = \begin{pmatrix} dz_{1,[0]} \\ \vdots \\ dz_{\gamma+1,[0]} \\ dz_{\gamma+2,[0]} \end{pmatrix} = \begin{pmatrix} d\lambda_0(\mathbf{x}_{[0]}) \\ \vdots \\ d\lambda_\gamma(\mathbf{x}_{[\gamma]}) \\ d\lambda_{\gamma+1}(\mathbf{x}) \end{pmatrix} = T(\mathbf{x}, \delta)d\mathbf{x}_{[0]} \tag{4.17}$$

is a bicausal change of coordinates.

  Consider $\dot{\lambda}_i(\mathbf{x})$ for $i \in [0, \gamma]$. By construction,

$$d\lambda_i(\mathbf{x})g_{1,j}(\mathbf{x}, \delta) = 0, \quad i \in [0, \gamma].$$

Consequently, if α is the maximum delay in $\lambda_i(\mathbf{x})$, that is, $\lambda_i := \lambda_i(\mathbf{x}_{[\alpha]})$, then

$$\dot{\lambda}_i(\mathbf{x}_{[\alpha]}) = \sum_{j=0}^{\alpha} \frac{\partial \lambda_i(\mathbf{x}_{[\alpha]})}{\partial x(-j)} F(\mathbf{x}(-j)), \quad i \in [0, \gamma].$$

Let $d\lambda_i(\mathbf{x}) = \Lambda_i(\mathbf{x}, \delta)d\mathbf{x}_{[0]}$, then

$$\begin{aligned} d\dot{\lambda}_i(\mathbf{x}) &= \dot{\Lambda}_i(\mathbf{x}, \delta)d\mathbf{x}_{[0]} + \Lambda_i(\mathbf{x}, \delta)d\dot{\mathbf{x}}_{[0]} \\ &= \dot{\Lambda}_i(\mathbf{x}, \delta)d\mathbf{x}_{[0]} + \Lambda_i(\mathbf{x}, \delta)f(\mathbf{x}, \mathbf{u}, \delta)d\mathbf{x}_{[0]} = \Gamma(\mathbf{x}, \delta)d\mathbf{x}_{[0]}. \end{aligned} \quad (4.18)$$

On the other hand, by assumption for any $k \geq 1$ and any $j \in [1, m]$,

$$\Lambda_i(\mathbf{x}, \delta)g_{k,j}(\mathbf{x}, \mathbf{u}, \delta) = 0,$$

so that derivating both sides, one gets $\forall k \geq 1, j \in [1, m]$,

$$\begin{aligned} 0 &= \dot{\Lambda}_i(\mathbf{x}, \delta)g_{k,j}(\mathbf{x}, \mathbf{u}, \delta) + \Lambda_i(\mathbf{x}, \delta)\dot{g}_{k,j}(\mathbf{x}, \mathbf{u}, \delta) \\ &= \Lambda_i(\mathbf{x}, \delta)g_{k,j}(\mathbf{x}, \mathbf{u}, \delta) + \Lambda_i(\mathbf{x}, \delta)f(\mathbf{x}, \mathbf{u}, \delta)g_{k,j}(\mathbf{x}, \mathbf{u}, \delta). \end{aligned} \quad (4.19)$$

It follows that for any $k \geq 1$ and any $j \in [1, m]$, by considering that $d\dot{\lambda}_i(\mathbf{x})$ is given by (4.18), then, due to (4.19),

$$\Gamma(\mathbf{x}, \delta)g_{k,j}(\mathbf{x}, \mathbf{u}, \delta) = \dot{\Lambda}_i(\mathbf{x}, \delta)g_{k,j}(\mathbf{x}, \mathbf{u}, \delta) + \Lambda_i(\mathbf{x}, \delta)f(\mathbf{x}, \mathbf{u}, \delta)g_{k,j}(\mathbf{x}, \mathbf{u}, \delta) = 0.$$

As a consequence, $d\dot{\lambda}_i \in \text{span}_{\mathcal{K}(\delta)}\{d\lambda_0(\mathbf{x}_{[0]}), \ldots, d\lambda_\gamma(\mathbf{x}_{[\gamma]})\}$ for any $i \in [0, \gamma]$. Accordingly, in the coordinates (4.17), the system necessarily reads (4.16).

Example 4.2 contin'd. Consider again system (4.2). Its differential representation is

$$d\dot{\mathbf{x}}_{[0]} = \begin{pmatrix} 0 & \mathbf{u}_{[0]}(-1)\delta \\ 0 & 0 \end{pmatrix} d\mathbf{x}_{[0]} + \begin{pmatrix} x_{2,[0]}(-1)\delta \\ 1 \end{pmatrix} d\mathbf{u}_{[0]}.$$

In this case, again through standard computations, one gets that

$$g_1(\mathbf{x}, \delta) = \begin{pmatrix} x_{2,[0]}(-1)\delta \\ 1 \end{pmatrix}, \quad g_2(\mathbf{x}, \mathbf{u}, \delta) = \begin{pmatrix} 0 \\ 0 \end{pmatrix}.$$

The accessibility matrix

$$\mathcal{R}_2(\mathbf{x}, \mathbf{u}, \delta) = \begin{pmatrix} x_{2,[0]}(-1)\delta & 0 \\ 1 & 0 \end{pmatrix}$$

has clearly rank 1, which according to Theorem 4.1 implies that the system is not completely accessible.

According to Proposition 4.2, there exists an autonomous function, which can be computed by considering the distributions $\mathcal{G}_i$. Starting from $\mathcal{G}_0$ we get

$$\mathcal{G}_0 = \text{span}_{\mathcal{K}(\delta)} \begin{pmatrix} \mathbf{g}^0 \ \mathbf{g}^1 \ 0 \\ 0 \ 0 \ I \end{pmatrix} = \text{span}_{\mathcal{K}(\delta)} \begin{pmatrix} 0 & x_{2,[0]}(-1) & 0 & 0 \\ 1 & 0 & 0 & 0 \\ 0 & 0 & 1 & 0 \\ 0 & 0 & 0 & 1 \end{pmatrix}.$$

Standard computations show that

$$\bar{\mathcal{G}}_0 = \text{span}_{\mathcal{K}(\delta)} \left\{ \frac{\partial}{\partial x_{[0]}}, \frac{\partial}{\partial x_{[0]}(-1)} \right\}.$$

Let us now consider

$$\mathcal{G}_1 = \text{span}_{\mathcal{K}(\delta)} \begin{pmatrix} \mathbf{g}^0 & \mathbf{g}^1 & 0 & 0 \\ 0 & \mathbf{g}^0(-1) & \mathbf{g}^1(-1) & 0 \\ 0 & 0 & 0 & I \end{pmatrix}$$

$$= \text{span}_{\mathcal{K}(\delta)} \begin{pmatrix} 0 & x_{2,[0]}(-1) & 0 & 0 & 0 \\ 1 & 0 & 0 & 0 & 0 \\ 0 & 0 & x_{2,[0]}(-2) & 0 & 0 \\ 0 & 1 & 0 & 0 & 0 \\ 0 & 0 & 0 & 1 & 0 \\ 0 & 0 & 0 & 0 & 1 \end{pmatrix}.$$

One thus gets that

$$\bar{\mathcal{G}}_1 = \text{span}_{\mathcal{K}(\delta)} \left\{ \frac{\partial}{\partial x_{2,[0]}}, x_{2,[0]}(-1)\frac{\partial}{\partial x_{1,[0]}} + \frac{\partial}{\partial x_{2,[0]}(-1)}, \frac{\partial}{\partial x_{1,[0]}(-1)}, \frac{\partial}{\partial x_{[0]}(-2)} \right\}.$$

4.3 On the Computation of the Accessibility Submodules

The accessibility generators $\mathbf{g}_i(\mathbf{x}, \mathbf{u}^{[i-2]}, \delta)$'s are strictly linked to the Polynomial Lie Bracket, thus generalizing to the delay context their definition in the nonlinear delay-free case.

In fact, starting from the dynamics (4.4), we can consider

$$\bar{F}(\mathbf{x}, \mathbf{u}, \epsilon) := \left(F(\mathbf{x}) + \sum_{i=0}^{l} \sum_{j=1}^{m} G_{ji}(\mathbf{x}) u_{[0],j}(-i) \right) \epsilon(0).$$

For a given vector $\tau(\mathbf{x}, \mathbf{u}, \delta)$, let us define

$$ad_{\bar{F}(\mathbf{x},\mathbf{u},1)}\tau(\mathbf{x}, \mathbf{u}, \delta) := ad_{\bar{F}(\mathbf{x},\mathbf{u},\epsilon)}\tau(\mathbf{x}, \mathbf{u}, \delta)|_{\epsilon(0)=1}$$

$$= \dot{\tau}(\mathbf{x}, \mathbf{u}, \delta) - f(\mathbf{x}, \mathbf{u}, \delta)\tau(\mathbf{x}, \mathbf{u}, \delta) \qquad (4.20)$$

and iteratively for any $i > 1$

$$ad^{i}_{\bar{F}(\mathbf{x},\mathbf{u},1)}\tau(\mathbf{x}, \mathbf{u}, \delta) = ad^{i-1}_{\bar{F}(\mathbf{x},\mathbf{u},1)}\left(ad_{\bar{F}(\mathbf{x},\mathbf{u},1)}\tau(\mathbf{x}, \mathbf{u}, \delta) \right).$$

Accordingly, the accessibility submodule generators, introduced in Márquez-Martínez (1999, 2000), defined (up to the sign) as

$$g_{i+1,j}(\mathbf{x}, \mathbf{u}^{[i-1]}, \delta) = \dot{g}_{i,j}(\mathbf{x}, \mathbf{u}^{[i-2]}, \delta) - f(\mathbf{x}, \mathbf{u}, \delta)g_{i,j}(\mathbf{x}, \mathbf{u}^{[i-2]}, \delta)$$

are then given by

$$g_{i+1,j}(\mathbf{x}, \mathbf{u}^{[i-1]}, \delta) = ad^{i}_{\bar{F}(\mathbf{x},\mathbf{u},1)}g_{1,j}(\mathbf{x}, \delta), \qquad (4.21)$$

which implies that they can be expressed in terms of Extended Lie Brackets. In fact, standard but tedious computations show that setting

$$\bar{F}_0(\mathbf{x}, \delta) = \sum_{j=0}^{ns} \bar{F}_0^j(\mathbf{x})\delta^j = \sum_{j=0}^{ns} F(\mathbf{x})\delta^j,$$

for $i \leq n$, and $g_{il}(\mathbf{x}, 0, \delta) = \sum_{p=0}^{is} g_{il}^p(\mathbf{x}, 0)\delta^p$, and denoting $g_{il}^p(\mathbf{x}, 0)$ with $g_{il}^p(0)$, then

$$g_{il}(\mathbf{x}, 0, \delta) = ad^{i-1}_{\bar{F}(\mathbf{x},0,1)}g_{1l}(\mathbf{x}, \delta) = \sum_{p=0}^{is}[\bar{F}_0^{is}, g_{i-1,l}^p(0)]_{E_0}\delta^p$$

$$= \sum_{p=0}^{is}[\bar{F}_0^{is}, \ldots, [\bar{F}_0^{is}, g_{1l}^p]_{E_{is}}]_{E_0}\delta^p.$$

Analogously, $g_{il}(\mathbf{x}, \mathbf{u}, \delta) = ad^{i-1}_{\bar{F}(\mathbf{x},\mathbf{u},1)}g_{1l}(\mathbf{x}, \delta)$ is given by

$$g_{il}(\mathbf{x}, \mathbf{u}, \delta) = g_{il}(\mathbf{x}, 0, \delta) +$$

$$+ \sum_{j=1}^{m} \sum_{q=0}^{i-2} \sum_{\mu=1}^{i-1-q} \sum_{k=-p-is}^{p+is} \sum_{\ell=0}^{is} \binom{i-1}{\mu+q} c_\mu^q [g_{\mu,j}^{k+\ell}(0), g_{i-\mu-q,l}^{\ell}(0)]_{E_0} \delta^\ell u_j^{(q)}(-k)$$

$$+ m_i(\mathbf{x}, \mathbf{u}^{[i-3]}, \delta), \tag{4.22}$$

where $c_\mu^0 = c_1^q = 1$, and for $\mu > 1$, $q > 0$, $c_\mu^q = c_{\mu-1}^q + c_\mu^{q-1}$, and $m_i(\mathbf{x}, \mathbf{u}^{[i-3]}, \delta)$ is given by the linear combination, through real coefficients, of terms of the form

$$\sum_\ell [g_{\mu_1,j_1}^{i_1+\ell}(0), \dots, [g_{\mu_\nu,j_\nu}^{i_\nu+\ell}(0), g_{i-q,l}^{\ell}(0)]_{E_{is}}]_{E_0} \delta^\ell \prod_{\mu=1}^{\nu} u_{j_\mu}^{(\ell_\mu)}(-i_\mu),$$

where $\nu \in [2, i-1]$, $j_\mu \in [1, m]$, $q = \sum_{k=1}^{\nu} \ell_k + \mu_k \le i - 1$.

The following result can be easily proven.

Proposition 4.3 *If for some coefficient $\alpha(\mathbf{x}, \mathbf{u}, \delta)$, $g_{i+1,j}(\cdot)\alpha(\mathbf{x}, \mathbf{u}, \delta) \in \mathcal{R}_i$, then $\forall k \ge 0$ there exist coefficients $\bar{\alpha}_k(\mathbf{x}, \mathbf{u}, \delta)$ such that $g_{i+k+1,j}(\cdot)\bar{\alpha}_k(\mathbf{x}, \mathbf{u}, \delta) \in \mathcal{R}_i$.*

The possibility of expressing the accessibility generators $g_{il}(\cdot, \delta)$ as a linear combination of Extended Lie Brackets of the g_{kj}'s for $j \in [1, l-1]$ allows to prove that when the dimension of the accessibility modules $\mathcal{R}_i$'s stabilizes, then it cannot grow, so that $\mathcal{R}_n$ has maximal dimension over $\mathcal{K}(\delta]$, and one can state the next result, whose proof can be found in Califano and Moog (2017) and which is also at the basis of Theorem 4.1.

Proposition 4.4 *Let $k = rank_{\mathcal{K}(\delta]}(\mathcal{R}_n(\mathbf{x}, \mathbf{u}, \delta))$ for almost all $\mathbf{u}$, and set*

$$\mathcal{R}_n(\mathbf{x}, 0, \delta) = span_{\mathcal{K}(\delta]}\{\mathbf{g}_1(\mathbf{x}, \delta), \dots, \mathbf{g}_n(\mathbf{x}, 0, \delta)\}.$$

Then $\bar{\mathcal{R}}_n(\mathbf{x}, 0, \delta)$, the involutive closure of $\mathcal{R}_n(\mathbf{x}, 0, \delta)$ has rank k.

Example 4.3 cont'd: For the dynamics (2.28), we have shown that the accessibility matrix has full rank 2 for $u \ne 0$ since it is given by

$$\mathcal{R}_2(\mathbf{x}, \mathbf{u}, \delta) = \begin{pmatrix} x_{2,[0]}(-1) \, \mathbf{u}_{[0]}(-1) - \mathbf{u}_{[0]}\delta \\ 1 & 0 \end{pmatrix}.$$

$\mathcal{R}_2(\mathbf{x}, 0, \delta)$ instead has dimension 1, while its involutive closure has again dimension 2. In fact, we have that, using the Polynomial Lie Bracket

$$\mathcal{R}_2(\mathbf{x}, 0, \delta) = \begin{pmatrix} x_{2,[0]}(-1) \\ 1 \end{pmatrix}, \qquad \bar{\mathcal{R}}_2(\mathbf{x}, 0, \delta) = \begin{pmatrix} x_{2,[0]}(-1) & 1 \\ 1 & 0 \end{pmatrix},$$

as expected.

4.4　On t-Accessibility of Time-Delay Systems

As already underlined through examples, even if a system admits some autonomous functions, it may be still possible to move from some initial state to a final state in an open subset of $I\!\!R^n$ though some time constraints on the trajectories have to be satisfied. This property is called $t - accessibility$.

Of course, if a system is accessible it will also be t-accessible, so that the problem arises when the submodule

$$\mathcal{R}_n(\mathbf{x}, \mathbf{u}, \delta) = \text{span}_{\mathcal{K}(\delta)}\{\mathbf{g}_1(x, \delta), \ldots, \mathbf{g}_n(\mathbf{x}, \mathbf{u}, \delta)\}$$

has rank $n - j < n$. In the general case, there exist j autonomous functions $\lambda_i(x_{[s]})$, $i \in [i, j]$, such that

$$\dot{\lambda}_i = \varphi_i(\lambda_1, \ldots, \lambda_j, \ldots, \lambda_1(-l), \ldots, \lambda_j(-l)), \quad i \in [1, j].$$

A first simple result can be immediately deduced when all the autonomous functions depend on non-delayed state variables only. In this case, the next result can be stated.

Theorem 4.4 *Consider the dynamics (4.4) and assume that the system is not fully accessible with* rank $\mathcal{R}_n(\mathbf{x}, \mathbf{u}, \delta) = n - j$, $j > 0$. *If* $\bar{\mathcal{G}}_0$ *given by (4.15) satisfies* dim $\bar{\mathcal{G}}_0^\perp = j$, *then the given system is also not t-accessible.*

Proof Clearly, if dim $\bar{\mathcal{G}}_0^\perp = j$ then all the independent autonomous functions of the system are delay free. Let $\lambda_i(x(t))$, $i \in [1, j]$ be such independent functions. Then under the delay-free change of coordinates

$$\begin{pmatrix} z_1(t) \\ \vdots \\ z_j(t) \\ \chi_1(t) \\ \vdots \\ \chi_{n-j}(t) \end{pmatrix} = \begin{pmatrix} \lambda_1(x(t)) \\ \vdots \\ \lambda_j(x(t)) \\ \varphi_1(x(t)) \\ \vdots \\ \varphi_{n-j}(x(t)) \end{pmatrix},$$

where the functions φ_i, $i \in [1, n - j]$, are chosen to define any basis completion, the given system reads

$$\dot{z}_{[0]} = \tilde{F}_z(z_{[s]})$$

$$\dot{\chi}_{[0]} = \tilde{F}_\chi(z_{[s]}, \chi_{[s]}) + \sum_{i=0}^{l} \tilde{G}_{i,\chi}(z_{[s]}, \chi_{[s]})u(-i). \tag{4.23}$$

Once the initial condition $\phi(t)$ over the interval $[-s\tau, 0)$ is fixed, the z-variables evolve along fixed trajectories defined by the initial condition, which proves that the system is not t-accessible.

The driftless case represents a particular case, since one has that if $\lambda(\mathbf{x})$ is an autonomous element, then $\dot{\lambda} = 0$. As a consequence in order to be t-accessible, the autonomous functions for a driftless system cannot depend on $x(t)$ only, as underlined hereafter.

Corollary 4.1 *The dynamics (4.4) with $F(\mathbf{x}_{[s]}) = 0$ is t-accessible only if $\bar{\mathcal{G}}_0^\perp = 0$.*

Things become more involved when only part of the autonomous functions are delay free. In this case, a necessary condition can be obtained by identifying if there is a subset of the autonomous functions, which is characterized by functions which are delay free and whose derivative falls in the same subset. To this end, the following algorithm was proposed in Gennari and Califano (2018).

Let $\Omega = span\left\{d\lambda_1(x(t)), \ldots, d\lambda_j(x(t))\right\}$.

Algorithm 4.1

Start
Set $\Omega_0 = \Omega$
Let $\Omega_1 = span\{\omega \in \Omega_0 : \dot{\omega} \in \Omega_0\}$
Let $k := 1$
For $k \leq dim(\Omega_0)$
      If $\Omega_k = \Omega_{k-1}$ **goto** *found*
      Else
          Set $\Omega_{k+1} = span\{\omega \in \Omega_k : \dot{\omega} \in \Omega_k\}$
          Set $k \leftarrow k + 1$
End
found
Set $\Omega = max_{Integrable}(\Omega_k)$
      If $dim(span\{\Omega_k\}) = dim(span\{\Omega\})$
      goto *close*
      Else goto *start*
close;
End

Proposition 4.5 *Let j be the dimension of Ω. Then Algorithm 4.1 ends after $k \leq j$ steps.*

Based on the previous algorithm the following result was also obtained.

Theorem 4.5 *Suppose that system (4.4) is not accessible. Let* rank $\mathcal{R}_n(\mathbf{x}, \mathbf{u}, \delta) = n - j$ *and accordingly $\mathcal{R}_n^\perp(\mathbf{x}, \mathbf{u}, \delta) = $ span$_{\mathcal{K}(\delta]}\left\{d\lambda_1(\cdot), \ldots, d\lambda_j(\cdot)\right\}$. Let the first $\bar{j} < j$ functions be independent of the delayed variables, that is, $\lambda_i = \lambda_i(x(t))$, $i \in [1, \bar{j}]$. If Algorithm 4.1 applied to $\Omega = \left\{d\lambda_1, \ldots, d\lambda_{\bar{j}}\right\}$ ends with Ω nonzero, then system (4.4) is not t-accessible.*

The proof is based on the consideration that the algorithm identifies a subset of the autonomous functions which depend only on $x(t)$ and identifies a non-accessible subsystem with special characteristics. Of course, it is not the maximal non-accessible subsystem in general. Furthermore, since the algorithm works on one-forms which are delay-free, the maximal integrable codistribution contained in Ω_k can be computed by using standard results.

Example 4.2 *contin'd.* Let us consider again the dynamics (4.2). We have already shown that the system is not fully accessible, but it was t-accessible. As a matter of fact $\bar{\mathcal{G}}_0$ has full rank, so that the autonomous function was obtained from $\bar{\mathcal{G}}_1$ and was $\lambda(\mathbf{x}) = x_1(t) - \frac{1}{2}x_2^2(t-1)$.

4.5 Problems

1. Given $\tau_1(\mathbf{x}_{[p,s]}, \delta)$, let

$$\tau_i(\mathbf{x}, \mathbf{u}, \delta) := ad_{\bar{F}(\mathbf{x},\mathbf{u},1)}^{i-1} \tau_1(\mathbf{x}_{[p,s]}, \delta), \qquad \text{for} \quad i > 1.$$

Show that the following result holds true:

Proposition 4.6. *Given* $\alpha(\mathbf{x}, \mathbf{u}, \delta)$,

$$ad_{\bar{F}(\mathbf{x},\mathbf{u},1)}(\tau_1(\mathbf{x}, \delta)\alpha) = \tau_2(\mathbf{x}, \mathbf{u}, \delta)\alpha + \tau_1(\mathbf{x}, \delta)\dot{\alpha}$$

and more generally

$$ad_{\bar{F}(\mathbf{x},\mathbf{u},1)}^k(\tau_1(\mathbf{x}, \delta)\alpha) = \sum_{j=0}^{k} \binom{k}{j} \tau_{k-j+1}(\mathbf{x}, \mathbf{u}, \delta)\alpha^{(j)}. \qquad (4.24)$$

2. DUPLICATION OF THE DYNAMICS.
   (a) Find an autonomous element for the following linear time-delay system:

$$\begin{cases} \dot{x}_1(t) = x_1(t-1) + u(t) \\ \dot{x}_2(t) = x_2(t-1) + u(t) \end{cases}.$$

   (b) Find an autonomous element for the following delay-free nonlinear system:

$$\begin{cases} \dot{x}_1(t) = x_1(t)u(t) \\ \dot{x}_2(t) = x_2(t)u(t) \end{cases}.$$

   (c) Show that the following nonlinear time-delay system is fully accessible or, equivalently, there is no autonomous element:

$$\begin{cases} \dot{x}_1(t) = x_1(t-1)u(t) \\ \dot{x}_2(t) = x_2(t-1)u(t) \end{cases}.$$

3. PRACTICAL CHECK OF ACCESSIBILITY.

Consider the system

$$\Sigma_0 : \begin{cases} \dot{x}_1(t) = x_2(t) + u(t) \\ \dot{x}_2(t) = x_3(t-\tau)u(t), \\ \dot{x}_3(t) = u(t) \end{cases}$$

where the initial condition for x_3 is some smooth function $c(t)$ defined for $t \in [-\tau, 0]$.

Checking accessibility amounts to compute combinations of state variables with infinite relative degree, i.e. which are not affected by the control input.

(a) Set $c(t) = x_3(t-\tau)$ which is considered as a time-varying parameter which is not affected by the control input:

$$\tilde{\Sigma}_0 : \begin{cases} \dot{x}_1(t) = x_2(t) + u(t) \\ \dot{x}_2(t) = c(t)u(t) \\ \dot{x}_3(t) = u(t) \end{cases}.$$

Compute all functions depending on $x(t)$ whose relative degree is larger than or equal to 2.

(b) For $t \geq \tau$, set $x_4(t) = x_3(t-\tau)$, $u(t-\tau) = v(t)$ and the extended system

$$\Sigma_1 : \begin{cases} \dot{x}_1(t) = x_2(t) + u(t) \\ \dot{x}_2(t) = x_4(t)u(t) \\ \dot{x}_3(t) = u(t) \\ \dot{x}_4(t) = v(t) \end{cases}.$$

Compute all functions depending on $x(t)$ whose relative degree is larger than or equal to 2 for the dynamics Σ_1. Check its accessibility.

Compare the number of autonomous elements of Σ_1 with the number of autonomous elements of $\tilde{\Sigma}_0$.

Conclude about the accessibility of Σ_0

4. Consider the system

$$\dot{x}(t) = f(x(t), x(t-1), \dots, x(t-k), u(t)),$$

where the initial condition for x is some smooth function $c_1(t)$ for $t \in [-1, 0]$, $c_2(t)$ for $t \in [-2, -1[, \dots, c_k(t)$ for $t \in [-k, -(k-1)[$. Define the system

$$\Sigma_0 : \dot{x}(t) = f(x(t), c_1, \dots, c_k, u(t)),$$

where $c_1(t), \dots, c_k(t)$ are time-varying parameters.

Any autonomous element for Σ_0 is also an autonomous element for the initial system, at least for $t \in [0, 1]$, i.e. it will not depend on the control input for $t \in [0, 1]$.

Now, denote $\xi(t) = x(t - 1)$ and define the extended system

$$
\begin{aligned}
\dot{x}(t) &= f(x(t), \xi(t), x(t - 2), \ldots, x(t - k), u(t)) \\
\dot{\xi}(t) &= f(\xi(t), x(t - 2), \ldots, x(t - k - 1), u(t - 1)).
\end{aligned}
\tag{4.25}
$$

The latter is well defined for $t \geq 1$. Define the system

$$
\Sigma_1 : \begin{cases} \dot{x}(t) = f(x(t), \xi(t), c_2, \ldots, c_k, u(t)) \\ \dot{\xi}(t) = f(\xi(t), c_2, \ldots, c_k, c_{k+1}, u(t - 1)) \end{cases}
$$

for some time-varying parameter $c_{k+1}(t)$.

(a) Compare the number of autonomous elements of Σ_1 with the number of autonomous elements of Σ_0.

Any autonomous element for Σ_1 is also an autonomous element for the initial system, at least for $t \in [0, 2]$, i.e. it will not depend on the control input for $t \in [0, 2]$.

More generally, consider the extended system. Now, denote $\xi(t) = x(t - 1)$ and define the extended system

$$
\begin{aligned}
\dot{x}(t) &= f(x(t), \xi_1(t), \xi_2(t), \ldots, \xi_k(t), u(t)) \\
\dot{\xi}_1(t) &= f(\xi_1(t), \xi_2(t), \ldots, \xi_k(t), x(t - k - 1), u(t - 1)) \\
&\;\;\vdots \\
\dot{\xi}_k(t) &= f(\xi_k(t), x(t - k - 1), \ldots, x(t - 2k, u(t - k)).
\end{aligned}
\tag{4.26}
$$

The latter is well defined for $t \geq k$. Define the system

$$
\Sigma_k : \begin{cases} \dot{x}(t) = f(x(t), \xi_1(t), \xi_2(t), \ldots, \xi_k(t), u(t)) \\ \dot{\xi}_1(t) = f(\xi_1(t), \xi_2(t), \ldots, \xi_k(t), c_{k+1}, u(t - 1)) \\ \;\;\vdots \\ \dot{\xi}_k(t) = f(\xi_k(t), c_{k+1}, \ldots, c_{2k}, u(t - k)) \end{cases}
$$

for some parameters $c_{k+1}, \ldots, c_{2k}$.

(b) Compare the number of autonomous elements of Σ_k with the number of autonomous elements of Σ_{k-1}.

Any autonomous element for Σ_k is also an autonomous element for the initial system, at least for $t \in [0, k]$, i.e. it will not depend on the control input for $t \in [0, k]$.

Chapter 5
Observability

Observability is an important characteristics of a system, which allows to address several problems when only the output function can be measured. Many authors have focused on the observer design problem starting from linear systems to nonlinear ones, adding then delays. Several differences already arise between continuous-time and discrete-time nonlinear delay-free systems (see, for example, Andrieu and Praly 2006; Besançon 1999; Bestle and Zeitz 1983; Califano et al. 2009; Krener and Isidori 1983; Xia and Gao 1989). When dealing with systems affected by delays additional features have to be taken into account. Some aspects are discussed in Anguelova and Wennberg (2010), Germani et al. (1996, 2002).

The observability property of a system is considered as the dual geometric aspect of accessibility. This is true in the linear and nonlinear delay-free case. However, when delays are present some differences arise as it has already been underlined in Sect. 1.4. In the present chapter, we will highlight the main aspects.

There are two main features, in particular, which should be examined. The first one stands in the fact that it is not, in general, possible to decompose the system into an observable subsystem and a nonobservable one as shown in Sect. 1.4, which is instead possible with respect to the accessibility property. The second feature is instead linked to the fact that the two notions of weak and strong observability introduced in the linear delay context are not sufficient when dealing with nonlinear systems. This becomes particularly clear when dealing with the realization problem. It is sufficient to note that, as shown in Halas and Anguelova (2013), a linear retarded-type single-input single-output (SISO), weakly observable linear time-delay system of order n

© The Author(s), under exclusive license to Springer Nature Switzerland AG 2021   75
C. Califano and C. H. Moog, *Nonlinear Time-Delay Systems*,
SpringerBriefs in Control, Automation and Robotics,
https://doi.org/10.1007/978-3-030-72026-1_5

$$\dot{x}(t) = \sum_{j=0}^{s} A_j x(t - j\tau) + \sum_{j=0}^{s} B_j u(t - j\tau)$$

$$y(t) = \sum_{j=0}^{s} C_j x(t - j\tau), \tag{5.1}$$

with τ being constant, always admits a strongly observable realization of the same order n, since the input–output equation is again of order n and of retarded type. Surprisingly, this is no more true in the nonlinear case as it will be discussed later on.

Let us now consider the nonlinear time-delay system

$$\Sigma : \begin{cases} \dot{\mathbf{x}}_{[0]} = F(\mathbf{x}_{[s]}) + \sum_{j=1}^{m} \sum_{i=0}^{l} G_{ji}(\mathbf{x}_{[s]})\mathbf{u}_{[0],j}(-i) \\ \mathbf{y}_{[0]} = H(\mathbf{x}_{[s]}) \end{cases}. \tag{5.2}$$

The following definition is in order (Califano and Moog 2020).

Definition 5.1 System (5.2) with $x(t) \in R^n$ is said to be *weakly observable* if, setting

$$\begin{pmatrix} dy \\ d\dot{y} \\ \vdots \\ dy^{(n-1)} \end{pmatrix} = \mathcal{O}(\cdot, \delta)dx + \mathcal{G}(\cdot, \delta) \begin{pmatrix} du \\ d\dot{u} \\ \vdots \\ du^{(n-2)} \end{pmatrix},$$

the observability matrix $\mathcal{O}(x, u, \ldots, u^{(n-2)}, \delta)$ has rank n over $\mathcal{K}(\delta]$.

System (5.2) is said to be *strongly observable* if the observability matrix $\mathcal{O}(x, u, \ldots, u^{(n-2)}, \delta)$ is unimodular, that is, it has an inverse polynomial matrix in δ.

Based on this definition let us examine the next example borrowed from Garcia-Ramirez et al. (2016a).

Example 5.1 Consider the system

$$\begin{cases} \dot{x}(t) = x(t - 1)u(t) \\ y(t) = x(t) + x(t - 1). \end{cases} \tag{5.3}$$

The observability matrix is obtained from dy and is given by

$$(dy) = (1 + \delta)dx = \mathcal{O}(x, \delta)dx,$$

which shows that according to Definition 5.1 the given system is weakly observable, but not strongly observable, since $1 + \delta$ does not have a polynomial inverse.

The main difference between the definition of strong and weak observability stands in the fact that the first one ensures that the state at time t can be reconstructed from the observation output and its first $n - 1$ time derivatives at time t. However, if one refers to the example, which shows that the system is not strongly observable, it is also immediate to notice that considering a larger number of derivatives allows to reconstruct the state for almost all input sequences. In fact, if one considers also the first-order derivative

$$\dot{y} = x(-1)u + x(-2)u(-1), \tag{5.4}$$

after standard computations one gets that $x(t)$ can be recovered as a function of the output, its first-order derivative and the input, whenever $u(t) \neq u(t - \tau)$. In fact, one gets that

$$x(0) = y(0) + \frac{\dot{y}(0) - u(-1)y(-1)}{u(-1) - u}. \tag{5.5}$$

It becomes then necessary to introduce the notion of *regular observability* in order to characterize this property, which will be further examined in Sect. 5.2.

5.1 Decomposing with Respect to Observability

While in the delay-free case it is always possible to decompose a nonlinear system into two subsystems, one completely observable and the other representing the unobservable part, this cannot be achieved, in general, in the delay case, due to the fact that the output function and its derivatives may not necessarily be suitable to define a bicausal change of coordinates. We try in this section to focus on the conditions under which such a decomposition exists. Before going into the details we first underline the conditions which determine if a system is weakly or strongly observable. The following results hold true (Califano and Moog 2020).

Proposition 5.1 *System (5.2) with $x(t) \in \mathbb{R}^n$ is weakly observable if and only if*

$$rank \, (\text{span}_{\mathcal{K}(\delta]}\{dx\} \cap \text{span}_{\mathcal{K}(\delta]}\{dy, \ldots, dy^{(n-1)}, du, \ldots du^{(n-2)}\}) = n.$$

Proposition 5.2 *System (5.2) with $x(t) \in \mathbb{R}^n$ is strongly observable if and only if*

$$span_{K(\delta]}\{dx\} \subset \text{span}_{\mathcal{K}(\delta]}\{dy, \ldots, dy^{(n-1)}, du, \ldots du^{(n-2)}\}.$$

Assume now that the system is neither strongly nor weakly observable. This means that there exists an index $k < n$ such that

$$\text{rank}_{\mathcal{K}(\delta]}\mathcal{O}(\mathbf{x}, \cdot, \delta) = k < n.$$

If such a conditions is verified then one may try to decompose the system into an observable and a non observable subsystem.

5.1.1   The Case of Autonomous Systems

Let us thus consider the autonomous system (that is, without control)

$$\dot{\mathbf{x}}_{[0]} = F(\mathbf{x}_{[s]})$$
$$y_{[0]} = h(\mathbf{x}),$$
(5.6)

and let us assume that for the given system the observability matrix has rank k. Then

$$T_1(\mathbf{x}, \delta)dx = \begin{pmatrix} dy(\mathbf{x}) \\ \vdots \\ dy^{(k-1)}(\mathbf{x}) \end{pmatrix}$$
(5.7)

has full row rank and one may wonder if the output and its derivatives could be used to define a bicausal change of coordinates. The following result holds true.

Proposition 5.3 *Let k be the rank of the observability matrix for the dynamics (5.6). Then there exists a basis completion $\chi = \psi(\mathbf{x})$ such that*

$$\begin{pmatrix} z_1 \\ \vdots \\ z_k \\ \chi \end{pmatrix} = \begin{pmatrix} y \\ \vdots \\ y^{(k-1)} \\ \psi \end{pmatrix}$$

is a bicausal change of coordinates, if and only if $T_1(\mathbf{x}, \delta)$ in (5.7) is closed and its right-annihilator is causal.

Example 5.2 Let us consider again the dynamics

$$\dot{x}_1 = 0$$
$$\dot{x}_2 = 0$$
$$y = x_1 x_2(-1) + x_2 x_1(-1).$$

Let us consider the observability matrix

$$\mathcal{O}(\mathbf{x}, \delta) = \begin{pmatrix} x_2(-1) + x_2\delta & x_1(-1) + x_1\delta \\ 0 & 0 \end{pmatrix}.$$

The matrix has clearly rank 1 so the system is not completely observable. Let us now check the right annihilator of dy to see if it can be used to define a bicausal change of coordinates. As a matter of fact, one easily gets that

$$\mathrm{ker}dy = \begin{pmatrix} -x_1(-1) - x_1(+1)\frac{x_2x_1(-2)-x_1x_2(-2)}{x_2(+1)x_1(-1)-x_2(-1)x_1(+1)}\delta \\ x_2(-1) + x_2(+1)\frac{x_2x_1(-2)-x_1x_2(-2)}{x_2(+1)x_1(-1)-x_2(-1)x_1(+1)}\delta \end{pmatrix},$$

which is not causal, and cannot be rendered causal. As a consequence, dy cannot be used to define a bicausal change of coordinates.

Example 5.3 Let us consider the dynamics

$$\dot{x}_1 = x_1(-1) + x_2(-1) - x_2(-2) + x_1(-1)x_2^2(-2) - x_2^3(-2)$$
$$\dot{x}_2 = x_1x_2^2(-1) + x_2 - x_2^3(-1)$$
$$y = x_1 - x_2(-1).$$

In this case, the observability matrix is

$$\mathcal{O}(\mathbf{x}, \delta) = \begin{pmatrix} 1 & -\delta \\ \delta & \delta^2 \end{pmatrix},$$

which has rank 1. dy is closed and

$$\mathrm{ker}dy = \begin{pmatrix} \delta \\ 1 \end{pmatrix}$$

which is causal. It can thus be used to define the desired bicausal change of coordinates. A possible solution is

$$\begin{pmatrix} z \\ \chi \end{pmatrix} = \begin{pmatrix} x_1 - x_2(-1) \\ x_2 \end{pmatrix}.$$

In the new coordinates, the given system reads

$$\dot{z} = z(-1)$$
$$\dot{\chi} = z\chi^2(-1) + \chi$$
$$y = z,$$

and we see that the system is split into an observable subsystem and an unobservable one.

5.2 On Regular Observability for Time-Delay Systems

As already underlined in the introductory section of this chapter, the regular observability notion is introduced to take into account the case in which the state can be reconstructed from the input and its derivatives up to an order which is greater than the state dimension. As it will be shown, this has important implications in the realization problem.

Definition 5.2 System (5.2) is said to be regularly observable if there exists an integer $N \geq n$ such that setting

$$
\begin{pmatrix} dy \\ d\dot{y} \\ \vdots \\ dy^{(N-1)} \end{pmatrix} = \mathcal{O}_e(\cdot, \delta)dx + \mathcal{G}_e(\cdot, \delta) \begin{pmatrix} du \\ d\dot{u} \\ \vdots \\ du^{(N-2)} \end{pmatrix}, \tag{5.8}
$$

the extended observability matrix $\mathcal{O}_e(x, u, \ldots, u^{(N-2)}, \delta)$ has rank n and admits a polynomial left inverse.

Let us underline that the previous definition of regular observability not only implies that the state can be recovered from the output, the input, and their derivatives, but also that the obtained function is causal. As an example consider the system

$$
\dot{x}(t) = u(t)
$$
$$
y(t) = x(t-1).
$$

Such a system is weakly observable, but not regularly observable, since $x(t) = y(t+1)$, which is not causal.

Note that strong observability implies regular observability and the latter yields weak observability.

Necessary and sufficient conditions can now be given to test regular observability. One has the following result whose proof is found in Califano and Moog (2020).

Proposition 5.4 *System (5.2) is regularly observable if and only if there exists an integer $N \geq n$ such that*

$$
\text{span}_{\mathcal{K}(\delta]}\{dx(t)\} \subset \text{span}_{\mathcal{K}(\delta]}\{dy(t), \ldots, dy^{(N-1)}(t), du(t), \ldots du^{(N-2)}(t)\}.
$$

Example 5.1 contin'd. For system (5.3), $N = 2$ since using Eqs. (5.3) and (5.4) yields

$$
\mathcal{O}_e(\cdot, \delta) = \begin{pmatrix} 1 + \delta \\ u\delta + u(-1)\delta^2 \end{pmatrix}
$$

and accordingly for $u \neq u(-1)$,

$$
T_0(\cdot, \delta) = \left(1 + \frac{u(-1)}{u-u(-1)}\delta \quad \frac{1}{u(-1)-u}\right),
$$

represents the left polynomial inverse of $\mathcal{O}_e$ thus proving that (5.3) is regularly observable as Proposition 5.4 holds true. In fact, the state at time t can be reconstructed through (5.5) whenever $u \neq u(-1)$. Of course, $u = u(-1)$ is a singular input. One can easily verify that the system is only weakly observable.

Finally, one may inquire on the difference between strong and regular observability. This is clarified by the following result (Califano and Moog 2020).

Corollary 5.1 *System (5.2) is regularly observable, and not strongly observable only if*

$$dy^{(n)} \notin \mathrm{span}_{\mathcal{K}(\delta]}\{dy, \ldots, dy^{(n-1)}, du, \ldots du^{(n-1)}\}.$$

The different properties of weak, strong, and regular observability have important implications on the input–output representation. In fact, it is easily verified that the following result holds true.

Theorem 5.1 *An observable SISO system with retarded-type state-space realization of order n admits a retarded-type input–output equation of order n if and only if*

$$dy^{(n)} \in \mathrm{span}_{\mathcal{K}(\delta]}\{dy, \ldots, dy^{(n-1)}, du, \ldots du^{(n-1)}\}, \tag{5.9}$$

that is, the system is strongly observable or it is weakly observable but not regularly observable.

Theorem 5.1 completes the statement of Theorem 2 in Califano and Moog (2020).

Proposition 5.5 follows from Theorem 5.1 and is peculiar of nonlinear time-delay systems as already anticipated.

Proposition 5.5 *A weakly observable SISO with retarded-type state-space realization of order n, such that*

$$dy^{(n)} \notin \mathrm{span}_{\mathcal{K}(\delta]}\{dy, \ldots, dy^{(n-1)}, du, \ldots du^{(n-1)}\}$$

admits a neutral-type input–output equation of order n and a retarded-type input–output equation of order $n + 1$ (and not smaller).

Example 5.1 contin'd. The dynamics (5.3) is strongly controllable for $x(-1) \neq 0$ and weakly observable. For the given retarded-type system of dimension $n = 1$, one has

$$dy = (1 + \delta)dx$$
$$d\dot{y} = (1 + \delta)u\delta dx + (1 + \delta)x(-1)du.$$

Clearly, $d\dot{y} \notin \mathrm{span}_{\mathcal{K}(\delta]}\{dx, du\}$, thus according to Proposition 5.5 it admits a neutral-type input–output equation of order one, which for $u \neq u(-1)$ is

$$(u - u(-1))\dot{y}(-1) - (u - u(-1))y(-2)u(-2)$$
$$+ (\dot{y} - y(-1)u)(u(-1) - u(-2)) = 0, \tag{5.10}$$

and a second-order retarded-type input–output equation given by

$$\ddot{y} = x(-1)\dot{u} + x(-2)\dot{u}(-1) + x(-2)u(-1)u + x(-3)u(-2)u(-1)$$
$$= y(-2)u(-1)u + (\dot{u}(-1) - \dot{u})\frac{\dot{y} - y(-1)u}{u(-1) - u}$$
$$+ y(-1)\dot{u} + u(-1)[u(-2) - u]\frac{\dot{y}(-1) - y(-2)u(-1)}{u(-2) - u(-1)}.$$

The previous example shows an important issue. In fact, one gets that to the one-dimensional retarded-type system (5.3) one can associate either a neutral-type first-order input–output equation or a second-order retarded-type input–output equation. In this second case, one has that using, for example, the procedure in Kaldmäe and Kotta (2018), one can compute a second-order retarded-type system associated to the input–output equation. To obtain it, one has to consider the filtration of submodules

$$\mathcal{H}_{i+1} = \{\omega \in \mathcal{H}_i | \dot{\omega} \in \mathcal{H}_i\}$$

with $\mathcal{H}_1 = \mathrm{span}_{\mathcal{K}(\delta]}\{dy^{(j)}, du^{(j)}, j \in [0, n-1]\}$, and check if $\mathcal{H}_{n+1}$ is integrable. In our case $n = 2$, and we then have to verify if $\mathcal{H}_3$ is integrable. In this case, for $u \neq u(-1)$

$$\mathcal{H}_1 = \mathrm{span}_{\mathcal{K}(\delta]}\{dy, d\dot{y}, du, d\dot{u}\}, \quad \mathcal{H}_2 = \mathrm{span}_{\mathcal{K}(\delta]}\left\{dy, d\left(\frac{\dot{y} - y(-1)}{u(-1) - u}\right), du\right\}$$
$$\mathcal{H}_3 = \mathrm{span}_{\mathcal{K}(\delta]}\left\{dy, d\left(\frac{\dot{y} - y(-1)}{u(-1) - u}\right)\right\}.$$

Since $\mathcal{H}_3$ is integrable, the state variable candidates result from its integration. So, we can set $x_1 = y, x_2 = \dfrac{\dot{y} - uy(-1)}{u(-1) - u}$, and we get the second-order retarded-type system

$$\begin{cases} \dot{x}_1 = x_1(-1)u + x_2[u(-1) - u] \\ \dot{x}_2 = x_2(-1)u(-2) \\ y = x_1. \end{cases} \tag{5.11}$$

The associated observability matrix is unimodular provided again that $u(-1) \neq u$, that is, the system is not solicited with a periodic input of period τ. As for the accessibility property, standard computations show that the system is weakly accessible.

We thus started with a one-dimensional retarded system strongly accessible for $x(-1) \neq 0$ and weakly observable for $u \neq u(-1)$, and we ended on a two-dimensional system which is weakly accessible and strongly observable, under appropriate conditions. This result which seems a contradiction is due to the fact that we are neglecting the relation established by the first-order neutral-type input–output equation, which links the state variables through the relation

$$p(x) = x_2 + x_2(-1) - x_1(-2) = 0. \tag{5.12}$$

If this condition is satisfied, while $u \neq 0$ and $\alpha \neq 0$, then the accessibility matrix has rank 1, and, after standard computations, one gets that $dp(\mathbf{x}) = (1 + \delta)dx_2 - \delta^2 dx_1$ is in the left annihilator of $\mathcal{R}$. Using the results of Chap. 4, we can then consider the bicausal change of coordinates $z_1 = x_2 + x_2(-1) - x_1(-2)$, $z_2 = x_1 - x_1(-1) + x_2$. In these coordinates, the system then reads

$$
\begin{aligned}
\dot{z}_1 &= z_1(-1)u(-2) \\
\dot{z}_2 &= (z_2(-1) - z_1)u + z_1 u(-1) \\
y &= z_2 + z_2(-1) - z_1
\end{aligned}
$$

and since by assumption $z_1 = 0$, we get

$$
\begin{aligned}
\dot{z}_1 &= 0 \\
\dot{z}_2 &= z_2(-1)u \\
y &= z_2 + z_2(-1),
\end{aligned}
$$

which highlights our weakly (and for $u \neq u(-1)$ also regularly) observable and strongly accessible subsystem of dimension one.

Further Note on the Observer Design

The transformation of the dynamics (5.2) by a change of coordinates and a nonlinear input–output injection is solved in Califano et al. (2013). These transformations allow to obtain delayed linear dynamics (up to nonlinear input–output injections) which are weakly observable. The design of a state observer can then be achieved using all the techniques available for linear time-delay systems (Garcia-Ramirez et al. 2016b).

5.3 Problems

1. Let the system

$$
\begin{cases}
\dot{x}_1(t) = x_2(t) + x_3(t)u(t) \\
\dot{x}_2(t) = x_1(t) \\
\dot{x}_3(t) = 0 \\
\dot{x}_4(t) = 0 \\
y(t) = x_1(t) + x_1(t-1)
\end{cases}.
$$

Check the strong, regular, and weak observability of the given system.

2. Consider the system

$$\begin{cases} \dot{x}_1(t) = u(t) \\ \dot{x}_2(t) = x_1(t) \\ \dot{x}_3(t) = 0 \\ y(t) = x_1(t)x_2(t-1) + x_1(t-1)x_2(t) \end{cases}$$

Compute $\dot{y}(t)$ and $\ddot{y}(t)$ and the Jacobian $\frac{\partial(y(t),\dot{y}(t),\ddot{y}(t))}{\partial(x_1(t),x_2(t),x_3(t))}$.
Is this system weakly observable?

Chapter 6
Applications of Integrability

In this chapter, it will be shown how the tools and methodologies introduced in the previous chapters can be used to address classical control problems, underlying also the main difference in the results with respect to the delay-free case.

Consider the nonlinear time-delay system

$$\dot{x}(t) = f(x(t-i), u(t-i); i = 0, \ldots, d_{max}), \tag{6.1}$$

where $x(t) \in \mathbb{R}^n$ and $u(t) \in \mathbb{R}^m$. Also, assume that the function f is meromorphic. To simplify the presentation, the following notation is used: $x(\cdot) := (x(t), x(t-1), \ldots)$. The notation $\varphi(x(\cdot))$ means that function φ can depend on $x(t), \ldots, x(t-i)$ for some finite $i \geq 0$. The same notation is used for other variables.

6.1 Characterization of the Chained Form with Delays

One classical model which is considered in the delay-free context is the chained form, which was first proposed by Murray and Sastry (1993), since it has been shown that many nonholonomic systems in mobile robotics, such as the unicycle, n-trailers, etc., can be converted into this form. It also turned out that some effective control strategies could be developed for these systems because of the special structure of the chained system. A characterization of its existence in the delay-free case is found in Sluis et al. (1994). We may thus be interested in understanding the characterization of such a form in the delay context. In particular, hereafter, we will define the conditions under which a two-input driftless time-delay nonlinear system of the form

$$\dot{\mathbf{x}}_{[0]} = \sum_{j=0}^{s} g_1^j(\mathbf{x})u_1(t-j) + \sum_{j=0}^{s} g_2^j(\mathbf{x})u_2(t-j) \tag{6.2}$$

can be put, under bicausal change of coordinates $\mathbf{z}_{[0]} = \phi(\mathbf{x}_{[\alpha]})$, in the following chained form:

$$\Sigma_c \begin{cases} \dot{z}_1(t) & = u_i(t) \\ \dot{z}_2(t) & = z_3(t-k_3)u_i(t-r_2) \\ \quad\vdots \\ \dot{z}_{n-1}(t) = z_n(t-k_n)u_i(t-r_{n-1}) \\ \dot{z}_n(t) & = u_j(t), \end{cases} \tag{6.3}$$

where k_ℓ and $r_{\ell-1}$, for $\ell \in [3, n]$, are non-negative integers, $i, j \in [1, 2]$, and $i \neq j$. While the results here recalled are taken from Califano et al. (2013) and Li et al. (2011), more general Goursat forms as well as the use of feedback laws will need instead further investigation. A first important result stated hereafter is that the chained form (6.3) is actually accessible, so that to investigate if a system can be put in such a form. A first issue that must be verified is accessibility.

Proposition 6.1 *The chained form with delays Σ_c is accessible for $u \neq 0$.*

The proof, which is left as an exercise, can be easily carried out by considering the accessibility matrix (Li et al. 2016).

Let us now consider the differential form representation of (6.2) which is equal to

$$d\dot{\mathbf{x}}_{[0]} = f(\mathbf{x}, \mathbf{u}, \delta)d\mathbf{x}_{[0]} + g_{11}(\mathbf{x}, \delta)du_{1,[0]} + g_{12}(\mathbf{x}, \delta)du_{2,[0]} \tag{6.4}$$

and let $g_{\ell,j}(\mathbf{x}, \cdot, \delta) = ad_{f(\mathbf{x}, \mathbf{u}, \delta)}^{\ell-1} g_{1,j}(\mathbf{x}, \delta)$. The following conditions are necessary.

Theorem 6.1 *System (6.2) can be transformed into the chained form (6.3) under bicausal change of coordinates $\mathbf{z}_{[0]} = \phi(\mathbf{x})$ only if there exists $j \in [1, 2]$ such that setting $u = 1 = const$, and denoting by $\mathbf{G}_{\ell,j}(\mathbf{x}, \epsilon)$ the image of $g_{\ell,j}(\mathbf{x}, \bar{\mathbf{u}}, \delta)$ when it acts on the function $\epsilon(t)$, then the following conditions are satisfied:*

1. $[\mathbf{G}_{\ell,j}(\mathbf{x}, \epsilon), g_{k,j}(\mathbf{x}, \delta)] = 0$, *for $k, \ell \in [1, n-1]$;*
2. $rank_{\mathcal{K}(\delta)} \mathcal{L}_{n-2} = n$, *where setting $u = 1$, the sequence $\mathcal{L}_0 \subset \mathcal{L}_1 \subset \cdots$ is defined as*

$$\mathcal{L}_0 = span_{\mathcal{K}(\delta)}\{g_{1,1}(\mathbf{x}, \delta), g_{1,2}(\mathbf{x}, \delta)\},$$
$$\mathcal{L}_{k+1} = \mathcal{L}_k + span_{\mathcal{K}(\delta)}\{g_{k+2,j}(\mathbf{x}, \bar{\mathbf{u}}, \delta)\}, \ k \geq 0, \ j \neq i.$$

The previous theorem, whose proof is found in Li et al. (2016), shows, in particular, that in the delay case only one of the input channels, the jth one, has to satisfy the straightening theorem, that is, $[\mathbf{G}_{1j}(\mathbf{x}, \epsilon), g_{1j}(\mathbf{x}, \delta)] = 0$, while the ith channel does not have to satisfy it, still allowing a solution to the problem. This makes itself an important difference with respect to the delay-free case.

Starting from Theorem 6.1, we are now ready to give necessary and sufficient conditions for the solvability of the problem.

Theorem 6.2 *System (6.2) can be transformed in the chained form (6.3) under bicausal change of coordinates* $z_{[0]} = \phi(x)$ *if only if the conditions of Theorem 6.1 are satisfied and additionally*

(a) *there exist integers* $0 = l_1 \leq l_2 \leq \cdots \leq l_{n-1}$ *such that for* $u = 1 = const.$,

$$g_{k,j}(\mathbf{x}, \bar{\mathbf{u}}, \delta) = \bar{g}_{k,j}(\mathbf{x}, \delta)\delta^{l_k} \text{ for } k \in [1, n-1] \tag{6.5}$$

while $g_{n,j}(\mathbf{x}, \bar{\mathbf{u}}, \delta) = 0$, *and the submodules* Δ_k, *defined as*

$$\Delta_\ell = \operatorname{span}_{\mathcal{K}(\delta]}\{\bar{g}_{1,j}(\mathbf{x}, \delta), \ldots, \bar{g}_{\ell,j}(\mathbf{x}, \delta)\}, \quad \ell \in [1, n-1]$$
$$\Delta_n = \operatorname{span}_{\mathcal{K}(\delta]}\{\bar{g}_{1,j}(\mathbf{x}, \delta), \ldots, \bar{g}_{n-1,j}(\mathbf{x}, \delta), g_{1,i}(\mathbf{x}, \delta)\},$$

are closed and locally of constant rank for $k \in [1, n]$;
(b) *denoting by* $\mathbf{G}_{1,i}(\mathbf{x}, \epsilon)$, *the image of* $g_{1,i}(\mathbf{x}, \delta)$ *acting on the function* $\epsilon(t)$, *then the Polynomial Lie Bracket*

$$[\mathbf{G}_{1,i}(\mathbf{x}, \epsilon), g_{1,i}(\mathbf{x}, \delta)] \in \Delta_{n-1}$$

for all possible values of $\epsilon(k)$;
(c) *let* $Q(\mathbf{x}, \epsilon)$ *be the image of* $[\mathbf{G}_{1,i}(\mathbf{x}, \epsilon), g_{1,i}(\mathbf{x}, \delta)]|_{\epsilon=\bar{\epsilon}}$, *then for* $\ell \in [1, n-1]$,

$$[Q(\mathbf{x}, \epsilon), g_{\ell,j}(\mathbf{x}, \delta] \in \operatorname{span}_{\mathcal{R}(\delta]}\{g_{\ell+2,j}(\mathbf{x}, \bar{\mathbf{u}}, \delta)\}.$$

While the proof, which uses essentially the properties of the Polynomial Lie Bracket, can be found in Califano et al. (2013), we prefer to discuss the interpretation of the conditions. More precisely condition (a) together with Theorem 6.1 guarantees that the Δ_ℓ's are involutive and that a coordinates change can be defined connected to them. Condition (b) is weaker than the straightening theorem expected in the delay-free case, but guarantees together with condition (c) the particular structure of the chained form with delays.

Example 6.1 (Califano et al. 2013) Consider the dynamics

$$\dot{x}_1(0) = u_1(0)$$
$$\dot{x}_2(0) = \left(x_3(-2) - x_2(-3) + x_1^2(-5)\right)u_1(-3) + 2x_1(-2)u_1(-2)$$
$$\dot{x}_3(0) = u_2(0) + \left(x_3(-3) - x_2(-4) + x_1^2(-6)\right)u_1(-4).$$

From the computation of its differential representation, we get that

$$g_{1,1} = \begin{pmatrix} 1 \\ \left(x_3(-2) - x_2(-3) + x_1^2(-5)\right)\delta^3 + 2x_1(-2)\delta^2 \\ \left(x_3(-3) - x_2(-4) + x_1^2(-6)\right)\delta^4 \end{pmatrix}, \; g_{1,2} = \begin{pmatrix} 0 \\ 0 \\ 1 \end{pmatrix}$$

$$f(\mathbf{x}, \mathbf{u}, \delta) = \begin{pmatrix} 0 & 0 & 0 \\ 2u_1(-3)x_1(-5)\delta^5 + 2u_1(-2)\delta^2 & -u_1(-3)\delta^3 & u_1(-3)\delta^2 \\ 2u_1(-4)x_1(-6)\delta^6 & -u_1(-4)\delta^4 & u_1(-4)\delta^3 \end{pmatrix}.$$

Accordingly, for the constant input $u = 1$

$$g_{2,2} = - \begin{pmatrix} 0 \\ \delta^2 \\ \delta^3 \end{pmatrix} = - \begin{pmatrix} 0 \\ 1 \\ \delta \end{pmatrix} \delta^2, \; g_{3,2} = 0.$$

We thus have that

$$\mathbf{G}_{1,2}(\mathbf{x}, \epsilon) = \begin{pmatrix} 0 \\ 0 \\ 1 \end{pmatrix} \epsilon(0), \; \mathbf{G}_{2,2}(\mathbf{x}, \epsilon) = - \begin{pmatrix} 0 \\ \epsilon(-2) \\ \epsilon(-3) \end{pmatrix},$$

whereas

$$\mathbf{G}_{1,1}(\mathbf{x}, \epsilon) = \begin{pmatrix} \epsilon(0) \\ \left(x_3(-2) - x_2(-3) + x_1^2(-5)\right)\epsilon(-3) + 2x_1(-2)\epsilon(-2) \\ \left(x_3(-3) - x_2(-4) + x_1^2(-6)\right)\epsilon(-4) \end{pmatrix}.$$

It is immediate to verify that $[\mathbf{G}_{\ell,2}(\mathbf{x}, \epsilon), g_{k,2}(\mathbf{x}, \delta)] = 0$ for $\ell, k \in [1, 2]$ and all the conditions of Theorem 6.1 are fulfilled so that one can check the remaining conditions of Theorem 6.2. We have already seen that $g_{2,2} = \bar{g}_{2,2}\delta^2$ and that $g_{3,2} = 0$. Since also $[\mathcal{G}_{1,1}(\mathbf{x}, \epsilon), g_{1,1}(\mathbf{x}, \delta)] = 0$, all the necessary and sufficient conditions are satisfied. The desired change of coordinates is then

$$\mathbf{z}_{[0]} = \begin{pmatrix} x_1 \\ x_2 - x_1^2(-2) \\ x_3 - x_2(-1) + x_1^2(-3) \end{pmatrix}.$$

In these new coordinates, the system reads

$$\dot{z}_{1,[0]} = u_1(0)$$
$$\dot{z}_{2,[0]} = z_3(-2)u_1(-3)$$
$$\dot{z}_{3,[0]} = u_2(0).$$

6.2 Input–Output Feedback Linearization

Consider the nonlinear single-input single-output (SISO) time-delay system

$$\begin{cases} \dot{x}(t) = f(x(t-i), u(t-i); i = 0, \ldots, d_{max}), \\ y(t) = h(x(t-i), u(t-i); i = 0, \ldots, d_{max}), \end{cases} \tag{6.6}$$

where both the input $u(t) \in \mathbb{R}^m$ and the output $y(t) \in \mathbb{R}^p$. A causal state feedback is sought

$$u(t) = \alpha(x(t-i), v(t-i); i = 0, \ldots, k)$$

for some k, so that the input–output relation from the new input $v(t)$ to the output $y(t)$ is linear (involving eventually some delays).

6.2.1 Introductory Examples

Consider the second-order system

$$\begin{cases} \dot{x}_1(t) = f_1(x(t), \ldots, x(t-k)) + u(t) \\ \dot{x}_2(t) = f_2(x(t-i), u(t-i); i = 0, \ldots, d_{max}) \\ y(t) = x_1(t). \end{cases} \tag{6.7}$$

Whatever the functions f_1 and f_2 are, it is always possible to cancel f_1 with a maximal loss of observability by the causal feedback $u(t) = -f_1(x(t), \ldots, x(t-k)) + v(t)$ to get the linear delay-free closed-loop system $\dot{y}(t) = v(t)$. This mimics what is done in the delay-free case and which is known as the computed torque method in robotics.

Consider now

$$\begin{cases} \dot{x}_1(t) = a_1 x_1(t) + a_2 x_1(t-1) + f_1(x(t-\delta), \ldots, x(t-k)) + u(t-\delta) \\ \dot{x}_2(t) = f_2(x(t-i), u(t-i); i = 0, \ldots, d_{max}) \\ y(t) = x_1(t), \end{cases} \tag{6.8}$$

where a_1 and a_2 are real numbers and δ is an integer greater than or equal to 2. Whatever the functions f_1 and f_2 are, it is still possible to cancel $f_1(\cdot)$ and still with a maximal loss of observability by the causal feedback $u(t) = -f_1(x(t), \ldots, x(t-k+\delta)) + v(t)$ to get the linear time-delay closed-loop system $\dot{y}(t) = a_1 y(t) + a_2 y(t-1) + v(t-\delta)$. This solution is different from the standard one known for delay-free nonlinear systems as not all terms in $\dot{y}(t)$ are cancelled out by feedback: the intrinsic linear terms do not need to be cancelled out.

Consider further

$$\begin{cases} \dot{x}_1(t) = a_1 x_1(t) + a_2 x_2(t-1) + f_1(x(t-\delta), \ldots, x(t-k)) + u(t-\delta) \\ \dot{x}_2(t) = a_3 x_1(t) + f_1(x(t-\delta-1), \ldots, x(t-k)) + u(t-\delta-1) \qquad (6.9) \\ y(t) = x_1(t), \end{cases}$$

where a_1, a_2, and a_3 are real numbers and $\delta \geq 2$ is an integer. Due to causality of the feedback, there is no way to cancel out the term $a_2 x_2(t-1)$. Thus, there will be no loss of observability under any causal state feedback. Still the causal feedback $u(t) = -f_1(x(t), \ldots, x(t-k+\delta)) + v(t)$ yields the linear time-delay closed-loop system $\ddot{y}(t) = a_1 \dot{y}(t) + a_2 a_3 y(t-1) + \dot{v}(t-\delta) + a_2 v(t-\delta-1)$.

6.2.2   Static Output Feedback Solutions

In this section, we are looking for a causal static output feedback

$$u(t) = \alpha((y(t-i), v(t-i); i = 0, \ldots, k)$$

and a bicausal change of coordinates $\mathbf{z}_{[0]} = \phi(\mathbf{x}_{[\alpha]})$ which transform the system (6.6) into

$$\dot{\mathbf{z}}_{[0]} = \sum_{j=0}^{\bar{s}} A_j \mathbf{z}_{[0]}(-j) + B\mathbf{v}_{[\bar{s}]}$$

$$\mathbf{y}_{[0]} = \sum_{j=k}^{\bar{s}+k} C_{j-k} \mathbf{z}_{[0]}(-j). \tag{6.10}$$

To solve this problem, it is necessary that the dynamics (6.6) are intrinsically linear up to nonlinear input–output injections $\psi(y(\cdot), u(\cdot))$ in the sense of Califano et al. (2013). Further conditions are, however, to be fulfilled so that the number of nonlinearities to be cancelled out is smaller than the number of available control variables. This is an algebraic condition which is formalized as follows.

The solvability of the nonlinear equations involving the input requires

$$\dim cls_{I\!R_{[s]}}\{dy, d\psi\} \leq m + p, \tag{6.11}$$

where $cls\{\cdot\}$ denotes the closure of a given submodule. Recall that the closure of a given submodule M of rank r is the largest submodule of rank r containing M according to Definition 1.2.

Consider, for instance,

$$M = span\{dx_1(t) + x_3(t)dx_2(t-1) + dx_1(t-2) + x_3(t-2)dx_2(t-3)\}$$

which has rank $r = 1$. Its closure N is spanned by $dx_1(t) + x_3(t)dx_2(t - 1)$ and thus has rank 1 as well. Clearly, $M \subset N$ but $N \not\subset M$.

The solvability of those equations in $u(t)$, say $\psi(y(\cdot), u(\cdot)) = v(t)$, does not yet ensure the solution is causal.

Causality of the static state feedback is obtained from the condition

$$cls_{\mathbb{R}_{[s]}}\{dy, d\psi\} = span_{\mathbb{R}_{[s]}}\{dy, d\varphi(u(t), y(t - j))\} \qquad (6.12)$$

so that $\{dy(t), d\varphi(u(t), y(-\tau_j))\}$ is a basis of the module $cls_{\mathbb{R}_{[s]}}\{dy, d\psi\}$, for a finite number of j's and with $\frac{\partial \varphi}{\partial u(t)} \neq 0$.

The above discussion is summarized as follows.

Theorem 6.3 *System (6.6) admits a static output feedback solution to the input–output linearization problem if and only if*

 (i) (6.6) is linearizable by input–output injections ψ,
(ii) and conditions (6.11) and (6.12) are satisfied.

The following illustrative examples are in order.

Example 6.2

$$\begin{cases} \dot{x}_1(t) = x_2(t - 3) - \sin x_1(t - 1) + u(t) \\ \dot{x}_2(t) = 2 \sin x_1(t - 1) - 2u(t) \\ y(t) = x_1(t). \end{cases} \qquad (6.13)$$

The static output feedback $u(t) = \sin y(t - 1) + v(t)$ yields the closed-loop system

$$\begin{cases} \dot{x}_1(t) = x_2(t - 3) + v(t) \\ \dot{x}_2(t) = -2v(t) \\ y(t) = x_1(t), \end{cases} \qquad (6.14)$$

which is a linear time-delay system. Note that system (6.13) is already in the form (6.10) and (6.11) is fulfilled as well with $\psi = u(t) - \sin y(t - 1)$.

Example 6.3 Consider now

$$\begin{cases} \dot{x}_1(t) = x_2(t - 3) - \sin x_1(t - 1) + u(t - 1) \\ \dot{x}_2(t) = 2 \sin x_1(t - 1) - 2u(t) \\ y(t) = x_1(t). \end{cases} \qquad (6.15)$$

Though the system (6.15) is in the form (6.10), two input–output injections are required $\psi_1 = u(t - 1) - \sin y(t - 1)$ and $\psi_2 = 2 \sin y(t - 1) - 2u(t)$. Thus, (6.11) is not fulfilled and there is no static output feedback solution to the input–output linearization problem.

Example 6.4 A third example is as follows:

$$\begin{cases} \dot{x}_1(t) = x_2(t-3) - \sin x_1(t-1) - 3\sin x_1(t-2) + u(t) + 3u(t-1) \\ \dot{x}_2(t) = 2\sin x_1(t-1) - 2u(t) \\ y(t) = x_1(t). \end{cases} \qquad (6.16)$$

System (6.16) is again in the form (6.10) thanks to the two input–output injections $\psi_1 = u(t) + 3u(t-1) - \sin y(t-1) - 3\sin y(t-2)$ and $\psi_2 = 2\sin y(t-1) - 2u(t)$. However,

$$cls_{\mathbb{R}_{[\delta]}}\{dy, d\psi_1, d\psi_2\} = span_{\mathbb{R}_{[\delta]}}\{dy, d[u(t) - \sin y(t-1)]\}$$

so that (6.11) is fulfilled and the linearizing output feedback solution is again $u(t) = \sin y(t-1) + v(t)$. The linear closed loop reads

$$\begin{cases} \dot{x}_1(t) = x_2(t-3) + v(t) + 3v(t-1) \\ \dot{x}_2(t) = -2v(t) \\ y(t) = x_1(t). \end{cases} \qquad (6.17)$$

Example 6.5 The next example is as follows:

$$\begin{cases} \dot{x}(t) = x(t-3) - \sin x_1(t-1) + u(t) + u(t-1) \\ y(t) = x_t(t). \end{cases} \qquad (6.18)$$

Condition (6.12) is not fulfilled and there is no static output feedback which cancels out the nonlinearity in the system.

6.2.3   Hybrid Output Feedback Solutions

Whenever the conditions above are not fulfilled, there is no static output feedback which is able to cancel out all nonlinearities. Nevertheless, a broader class of output feedbacks may still offer a solution (Márquez-Martínez and Moog 2004).

Consider, for instance, the system

$$\begin{cases} \dot{x}(t) = -\sin x(t-2) + u(t) + u(t-1) \\ y(t) = x(t). \end{cases}$$

Condition (6.12) is not fulfilled as φ involves the control variable at different time instants. Previous elementary tools suggest to implement the implicit control law $u(t) = -u(t-1) + \sin y(t-2) + v(t)$ which would just yield $\dot{y}(t) = v(t)$. A practical realization of this control law is obtained setting $z(t) = u(t-1)$ in the form of an hybrid output feedback since it also involves the shift operator. Note that $z(t) = u(t-1)$ is nothing but a buffer. The resulting compensator is named a pure shift dynamic output feedback in Márquez-Martínez and Moog (2004) and for the example above it reads as

$$\begin{cases} z(t+1) = -z(t) + \sin y(t-2) + v(t) \\ \quad u(t) = -z(t) + \sin y(t-2) + v(t), \end{cases}$$

which yields $\dot{y}(t+1) = v(t+1)$, i.e. linearity of the input–output relation is obtained after some time only.

Accepting this broader class of output feedbacks allows to weaken the conditions in Theorem 6.4.

Theorem 6.4 *System (6.6) admits an hybrid output feedback solution to the input–output linearization problem if and only if*

(i) *(6.6) is linearizable by input–output injections;*
(ii) *condition (6.11) is satisfied;*
(iii) *and $cls_{\mathbb{R}_{[\delta]}}\{dy, d\psi\} = span_{\mathbb{R}_{[\delta]}}\{dy, d\varphi(u(t), \ldots, u(t-i), y(t-j))\}$, so that $\{dy(t), d\varphi(u(t), \ldots, u(t-i), y(t-j))\}$ is a basis of the module $cls_{\mathbb{R}_{[\delta]}}\{dy(t), d\psi\}$, for a finite number of j's and with $\frac{\partial \varphi}{\partial u(t)} \neq 0$.*

Note that the digital implementation of the stabilizing control in Monaco et al. (2017) goes through a trick similar to a buffer equation (see Eq. (18) in Monaco et al. 2017).

Further solutions to input–output feedback linearization including causal static state feedbacks and causal dynamic feedbacks are available in the literature (Germani et al. 1996; Márquez-Martínez and Moog 2004; Oguchi et al. 2002) and weaken the conditions of solvability of the problem.

6.3 Input-State Linearization

In general, input–output linearization just partially linearizes the system. This is known to be a drawback for delay-free systems which have some unstable zero dynamics. The situation is obviously the same in the case of nonlinear time-delay systems.

In such a case, the input to state feedback linearization, or full linearization, is specially helpful because the full state variables remain "under control" in the closed loop, and can be further controlled using more standard linear techniques available for linear time-delay systems. Note that full linearization via dynamic state feedback was popularized and is also known as flatness of control systems.

The input to state linearization via static state feedback is stated as follows. Given

$$\dot{x}(t) = f(x(t-i), u(t-i); i = 0, \ldots, d_{max}) \qquad (6.19)$$

find, if possible, a causal state feedback $u(t) = \alpha(x(t-i), v(t-i); i = 0, \ldots, k)$, a bicausal change of coordinates $z(t) = \varphi(x(t-i); i = 0, \ldots, \ell)$ and $K \in \mathbb{N}$ such that the dynamics is linear in the z coordinates after some time K:

$$\dot{z}(t) = A(\delta)z + B(\delta)v$$

for all $t \geq K$ and where the pair $A(\delta)$, $B(\delta)$ is weakly controllable.

6.3.1 Introductory Example

Consider the system

$$\begin{cases} \dot{x}_1(t) = x_2(t-1) \\ \dot{x}_2(t) = 3x_1(t-2) + \cos{(x_1^2(t-2))} + x_2(t-1)u(t-1). \end{cases}$$

Pick the "output function" $y(t) = x_1(t)$ which has a full relative degree equal to 2. A standard method valid for delay-free systems consists in cancelling by feedback the full right-hand side of $\ddot{y}(t)$. Due to the causality constraint this cannot be done in this special case. Nevertheless, one takes advantage of some pre-existing suitable structure and cancel only nonlinear terms. The obvious solution is

$$u(t) = \frac{v(t) - \cos{(x_1^2(t-1))}}{x_2(t)},$$

and the closed-loop system reads

$$\dot{x}(t) = \begin{bmatrix} 0 & \delta \\ 3\delta^2 & 0 \end{bmatrix} x(t) + \begin{bmatrix} 0 \\ \delta \end{bmatrix} v(t),$$

which is a weakly controllable linear system.

6.3.2 Solution

Focus on the single-input case (Baibeche and Moog 2016). Following the methodology valid for delay-free systems, one has to search for an output function candidate which has full relative degree, i.e. equal to n. Then, apply the results valid for input–output linearization. Such an output function is called a "linearizing output".

   This methodology will, however, yield conditions which are only sufficient for feedback linearization in the case of time-delay nonlinear systems.

   Nevertheless, a necessary and sufficient condition for the existence of a full relative degree output candidate can be derived thanks to the Polynomial Lie Bracket.

Lemma 6.1 *There exists $h(x(\cdot))$ with full relative degree n if and only if*

(i) rank $\bar{\mathcal{R}}_{n-1}(\mathbf{x}, 0, \delta) = n - 1$,
(ii) rank $\bar{\mathcal{R}}_n(\mathbf{x}, 0, \delta) = n$,

where $\mathcal{R}_i$ is defined by (4.8).

Condition (ii) ensures the accessibility of the system under interest, so that (i) yields the existence of an output function candidate with finite relative degree n.

Theorem 6.5 *Given system (6.19), there exists a causal static state feedback which solves the input-state linearization problem if the conditions of Lemma 6.1 are fulfilled and for some $dh \perp \bar{R}_{n-1}(\mathbf{x}, 0, \delta)$*

(i) $dh^{(n)} \in \mathrm{span}_{I\!R[\delta]}\{d[a(x(\cdot)) + b(x(\cdot))u(t-i)]\}$ *for some $i \in I\!N$ and $a, b \in \mathcal{K}$;*

(ii) $\frac{\partial[a(x(\cdot)) + b(x(\cdot))u(t-i)]}{\partial x(t-\tau)} = 0,\ 0 \leq \tau < i;$ *and*

(iii) *there exists a polynomial matrix $D(\delta)$ in $I\!R[\delta]^{n \times n}$ such that*

$$\begin{pmatrix} dh \\ d\dot{h} \\ \vdots \\ dh^{(n-1)} \end{pmatrix} = D(\delta)M(\mathbf{x}, \delta)dx$$

for some unimodular matrix $M(\mathbf{x}, \delta)$.

Note that condition (i) in Theorem 6.5 restricts the class of systems under interest as the control variable is subject to one single delay i. Then, condition (ii) ensures that the linearizing feedback

$$u(t) = \frac{v(t) - a(x(\cdot))}{b(x(\cdot))}$$

is causal. The closed loop then yields a linear time-delay differential input–output equation.

6.4 Normal Form

In this section, with reference to the SISO case, we are seeking for a bicausal change of coordinates $z(t) = \varphi(x(\cdot))$ and a regular causal feedback $u(t) = \alpha(x(\cdot), v(\cdot))$, which allows to transform the given dynamics (6.1) with output $y = h(\mathbf{x})$ into the form

$$\dot{z}^1(t) = \sum_{j=0}^{\bar{s}} A_j z^1(-j) + \sum_{j=0}^{\bar{s}} B_j v(-j)$$

$$\dot{z}^2(t) = f_2(z^1(\cdot), z^2(\cdot)), \tag{6.20}$$

$$y = \sum_{j=0}^{\bar{s}} C_j z^1(-j),$$

where the subsystem defined by the z^1 variable is linear, weakly or strongly observable, and weakly or strongly accessible.

As a matter of fact, one has to find a bicausal change of coordinates and a regular
and causal static feedback such that the input–output behavior is linearized and the
residual dynamics does not depend on the control.

The solution requires thus, on one hand, that the input–output linearization can
be achieved via a bicausal change of coordinates and a causal feedback, on the other,
that one can choose the basis completion to guarantee that the residual dynamics is
independent of the control. This can be formulated by checking some conditions on
the Polynomial Lie Bracket $[G_1(\mathbf{x}, \epsilon), g_1(\mathbf{x}, \delta)]$. While the result can be found in the
literature (Califano and Moog 2016), we leave the formulation of the solution and
its proof as an exercise for the reader.

6.5 Problems

1. Show that system (6.3) is accessible.
2. Consider the system

$$\begin{cases} \dot{x}_1(t) = x_2(t-1) \\ \dot{x}_2(t) = \cos(x_1^2(t-1)) + x_2(t-1)u(t). \end{cases}$$

Compute a linearizing output, if any, and the corresponding linearizing state feedback.
3. Check whether there exists an output function candidate for the following system:

$$\begin{cases} \dot{x}_1(t) = x_2(t-1) + 3[\cos(x_1^2(t-2)) + x_2(t-2)u(t-1)] \\ \dot{x}_2(t) = x_3(t) + \cos(x_1^2(t-1)) + x_2(t-1)u(t) \\ \dot{x}_3(t) = 2[\cos(x_1^2(t-3)) + x_2(t-3)u(t-2)]. \end{cases}$$

Compute an output function candidate whose relative degree equals 3, if any.
Compute the closed-loop system under the action of the static state feedback

$$u(t) = -\frac{\cos(x_1^2(t-1))}{x_2(t-1)}.$$

Check its accessibility.
4. Let the system

$$\begin{cases} \dot{x}_1(t) = x_2(t) + u(t-1)\sin x_1(t-2) \\ \dot{x}_2(t) = x_1(t) + u(t)\sin x_1(t-1) \\ y(t) = x_1(t). \end{cases}$$

Compute a causal static output feedback, if any, which linearizes the input–output
system.
Write the closed-loop system and check its weak accessibility.

5. Let the system

$$\begin{cases} \dot{x}_1(t) = x_2(t-2) + u(t-1)\sin x_1(t-2) \\ \dot{x}_2(t) = x_1(t) + u(t)\sin x_1(t-1) \\ \quad y(t) = x_1(t). \end{cases}$$

Compute a causal static output feedback, if any, which linearizes the input–output system.

Write the closed-loop system and check its weak accessibility.

Series Editor Biographies

Tamer Başar is with the University of Illinois at Urbana-Champaign, where he holds the academic positions of Swanlund Endowed Chair, Center for Advanced Study (CAS) Professor of Electrical and Computer Engineering, Professor at the Coordinated Science Laboratory, Professor at the Information Trust Institute, and Affiliate Professor of Mechanical Science and Engineering. He is also the Director of the Center for Advanced Study—a position he has been holding since 2014. At Illinois, he has also served as Interim Dean of Engineering (2018) and Interim Director of the Beckman Institute for Advanced Science and Technology (2008–2010). He received the B.S.E.E. degree from Robert College, Istanbul, and the M.S., M.Phil., and Ph.D. degrees from Yale University. He has published extensively in systems, control, communications, networks, optimization, learning, and dynamic games, including books on non-cooperative dynamic game theory, robust control, network security, wireless and communication networks, and stochastic networks, and has current research interests that address fundamental issues in these areas along with applications in multi-agent systems, energy systems, social networks, cyber-physical systems, and pricing in networks.

In addition to his editorial involvement with these Briefs, Başar is also the Editor of two Birkhäuser series on *Systems & Control: Foundations & Applications* and *Static & Dynamic Game Theory: Foundations & Applications*, the Managing Editor of the *Annals of the International Society of Dynamic Games* (ISDG), and member of editorial and advisory boards of several international journals in control, wireless networks, and applied mathematics. Notably, he was also the Editor-in-Chief of *Automatica* between 2004 and 2014. He has received several awards and recognitions over the years, among which are the Medal of Science of Turkey (1993); Bode Lecture Prize (2004) of IEEE CSS; Quazza Medal (2005) of IFAC; Bellman Control Heritage Award (2006) of AACC; Isaacs Award (2010) of ISDG; Control Systems Technical Field Award of IEEE (2014); and a number of international honorary doctorates and professorships. He is a member of the US National Academy of Engineering, a Life Fellow of IEEE, Fellow of IFAC, and Fellow of SIAM. He has served as an IFAC Advisor (2017-), a Council Member of IFAC (2011–2014), President of

© The Author(s), under exclusive license to Springer Nature Switzerland AG 2021
C. Califano and C. H. Moog, *Nonlinear Time-Delay Systems*,
SpringerBriefs in Control, Automation and Robotics,
https://doi.org/10.1007/978-3-030-72026-1

AACC (2010–2011), President of CSS (2000), and Founding President of ISDG (1990–1994).

Miroslav Krstic is Distinguished Professor of Mechanical and Aerospace Engineering, holds the Alspach Endowed Chair, and is the Founding Director of the Cymer Center for Control Systems and Dynamics at UC San Diego. He also serves as Senior Associate Vice Chancellor for Research at UCSD. As a graduate student, he won the UC Santa Barbara best dissertation award and student best paper awards at CDC and ACC. He has been elected as Fellow of IEEE, IFAC, ASME, SIAM, AAAS, IET (UK), AIAA (Assoc. Fellow), and as a foreign member of the Serbian Academy of Sciences and Arts and of the Academy of Engineering of Serbia. He has received the SIAM Reid Prize, ASME Oldenburger Medal, Nyquist Lecture Prize, Paynter Outstanding Investigator Award, Ragazzini Education Award, IFAC Nonlinear Control Systems Award, Chestnut textbook prize, Control Systems Society Distinguished Member Award, the PECASE, NSF Career, and ONR Young Investigator awards, the Schuck ('96 and '19) and Axelby paper prizes, and the first UCSD Research Award given to an engineer. He has also been awarded the Springer Visiting Professorship at UC Berkeley, the Distinguished Visiting Fellowship of the Royal Academy of Engineering, and the Invitation Fellowship of the Japan Society for the Promotion of Science. He serves as Editor-in-Chief of *Systems & Control Letters* and has been serving as Senior Editor for *Automatica and IEEE Transactions on Automatic Control*, as Editor of two Springer book series—*Communications and Control Engineering and SpringerBriefs in Control, Automation and Robotics*—and has served as Vice President for Technical Activities of the IEEE Control Systems Society and as Chair of the IEEE CSS Fellow Committee. He has coauthored 13 books on adaptive, nonlinear, and stochastic control, extremum seeking, control of PDE systems including turbulent flows, and control of delay systems.

References

Alvarez-Aguirre A, van de Wouw N, Oguchi T, Kojima K, Nijmeijer H (2011) Remote tracking control of unicycle robots with network-induced delays. In: Cetto JA et al (eds) Informatics in Control, Automation and Robotics. LNEE 89. Springer, New York, pp 225–238

Andrieu V, Praly L (2006) On the existence of a Kazantzis-Kravaris/Luenberger observer. SIAM J Cont Opt 45(2):432–456

Anguelova M, Wennberg B (2010) On analytic and algebraic observability of nonlinear delay systems. Automatica 46:682–686

Baibeche K, Moog CH (2016) Input-state feedback linearization for a class of single-input nonlinear time-delay systems. J Math Control Inf 33:873–891

Banks PS (2002) Nonlinear delay systems, Lie algebras and Lyapunov transformations. J Math Control Inf 19:59–72

Battilotti S (2020) Continuous-time and sampled-data stabilizers for nonlinear systems with input and measurement delays. IEEE Trans Autom Control 65:1568–1583

Besançon G (1999) On output transformations for state linearization up to output injection. IEEE Trans Autom Control 44:1975–1981

Bestle D, Zeitz M (1983) Canonical form observer design for non-linear time-variable systems. Int J Control 38:419–431

Bloch M (2003) Nonholonomic mechanics and control. Springer, New York

Buckalo AF (1968) Explicit conditions for controllability of linear systems with time lag. IEEE Trans Autom Control 13:193–195

Califano C, Li S, Moog CH (2013) Controllability of driftless nonlinear time-delay systems. Syst Contr Lett 62:294–301. https://doi.org/10.1016/j.sysconle.2012.11.023

Califano C, Márquez-Martínez LA, Moog CH (2010) On linear equivalence for time–delay systems. In: Proceedings of the 2010 American control confernce, art no 5531404, pp 6567–6572, Baltimore, USA

Califano C, Márquez-Martínez LA, Moog CH (2011) Extended Lie brackets for nonlinear time-delay systems. IEEE Trans Automat Control 56:2213–2218. https://doi.org/10.1109/TAC.2011.2157405

Califano C, Márquez-Martínez LA, Moog CH (2011) On the observer canonical form for time–delay systems. In: Proceedings of 18th Ifac world congress 2011, V 18, Part 1, Milan, Italy, pp 3855–3860. https://doi.org/10.3182/20110828-6-IT-1002.00729

Califano C, Márquez-Martínez LA, Moog CH (2013) Linearization of time-delay systems by input-output injection and output transformation. Automatica 49:1932–1940

© The Author(s), under exclusive license to Springer Nature Switzerland AG 2021
C. Califano and C. H. Moog, *Nonlinear Time-Delay Systems*,
SpringerBriefs in Control, Automation and Robotics,
https://doi.org/10.1007/978-3-030-72026-1

Califano C, Monaco S, Normand-Cyrot D (2009) Canonical observer forms for multi-output systems up to coordinate and output transformations in discrete time. Automatica 45:2483–2490. https://doi.org/10.1016/j.automatica.2009.07.003

Califano C, Moog CH (2014) Coordinates transformations in nonlinear time-delay systems. In: 53rd IEEE conference on decision and control. Los Angeles, USA, pp 475–480

Califano C, Moog CH (2016) On the existence of the normal form for nonlinear delay systems. In: Karafyllis I, Malisoff M, Mazenc F, Pepe P (eds) Recent results on nonlinear delay control systems, vol 4. Advances in delays and dynamics. Springer, Berlin Heidelberg New York, pp 113–142

Califano C, Moog CH (2017) Accessibility of nonlinear time-delay systems. IEEE Trans Autom Control 62:7. https://doi.org/10.1109/TAC.2016.2581701

Califano C, Moog CH (2020) Observability of nonlinear time-delay systems and its application to their state realization. IEEE Control Syst Lett 4: 803–808, https://doi.org/10.1109/LCSYS.2020.2992715

Choquet-Bruhat Y, DeWitt-Morette C, Dillard-Bleick M (1989) Analysis, manifolds and physics, part I: basics. North-Holland, Amsterdam

Cohn PM (1985) Free rings and their relations. Academic Press, London

Conte G, Moog CH (2007) Perdon A (2007) Algebraic methods for nonlinear control systems, 2nd edn. Springer, London, p

Conte G, Perdon AM (1995) The disturbance decoupling problem for systems over a ring. SIAM, J Contr Optimiz 33:750–764

Crouch PE, Lamnabhi-Lagarrigue F, Pinchon D (1995) Some realizations of algorithms for nonlinear input-output systems. Int J Contr 62:941–960

Fliess M, Mounier H (1998) Controllability and observability of linear delay systems: an algebraic approach. ESAIM COCV 3:301–314

Fridman E (2001) New Lyapunov-Krasovskii functionals for stability of linear retarded and neutral type systems. Syst Control Lett 43:309–319

Fridman E (2014) Introduction to time-delay systems. Birkhäuser, Basel

Fridman E, Shaked U (2002) A descriptor system approach to H-infinity control of linear time-delay systems. IEEE Trans Aut Contr 47:253–270

Garcia-Ramirez E, Moog CH, Califano C (2016) LA Marquez-Martinez Linearisation via input-output injection of time delay systems. Int J Control 89:1125–1136

Garcia-Ramirez E, Califano C, Marquez-Martinez LA, Moog CH (2016) Observer design based on linearization via input-output injection of time-delay systems. Proceedings IFAC NOLCOS, Monterey, CA, USA, IFAC-PapersOnLine 49–18(2016):672–677

Gennari F, Califano C (2018) T-accessibility for nonlinear time-delay systems: the general case. In: IEEE conference on decision and control (CDC), pp 2950–2955

Germani A, Manes C, Pepe P (1996) Linearization of input-output mapping for nonlinear delay systems via static state feedback. In: CESA '96 IMACS multiconference, pp 599–602

Germani A, Manes C, Pepe P (1998) Linearization and decoupling of nonlinear delay systems. In: Proceedings of the American control conference, Philadelphia, USA, pp 1948–1952

Germani A, Manes C, Pepe P (2002) A new approach to state observation of nonlinear systems with delayed output. IEEE Trans Autom Control 47:96–101

Glumineau A, Moog CH, Plestan F (1996) New algebro-geometric conditions for the linearization by input-output injection. IEEE Trans Autom Control 41:598–603

Gu K, Kharitonov VL, Chen J (2003) Stability of time-delay systems. Birkhaüser, Boston

Hermann R (1963) On the accessibility problem in control theory. In: Lasalle JP, Efschetz SL (eds) International symposium on nonlinear differential equations and nonlinear mechanics. Academic Press, pp 325–332

Halas M, Anguelova M (2013) When retarded nonlinear time-delay systems admit an input-output representation of neutral type. Automatica 49:561–567

Hespanha JP, Naghshtabrizi P, Xu Y (2007) A survey of recent results in networked control systems. Proc IEEE 95:138–162

Hou M, Pugh AC (1999) Observer with linear error dynamics for nonlinear multi-output systems. Syst Contr Lett 37:1–9

Insperger T (2015) On the approximation of delayed systems by Taylor series expansion. ASME J Comput Nonlinear Dyn 10:1–4

Isidori A (1995) Nonlinear control systems, 3rd edn. Springer, New York

Islam S, Liu XP, El Saddik A (2013) Teleoperation systems with symmetric and unsymmetric time varying communication delay. IEEE Trans Instrum Meas 21:40–51

Kaldmäe A, Kotta Ü (2018) Realization of time-delay systems. Automatica 90:317–320

Kaldmäe A, Moog CH, Califano C (2015) Towards integrability for nonlinear time-delay systems. In: MICNON 2015. St Petersburg, Russia, IFAC-PapersOnLine 48, pp 900–905. https://doi.org/10.1016/j.ifacol.2015.09.305

Kaldmäe A, Califano C, Moog CH (2016) Integrability for nonlinear time-delay systems. IEEE Trans Autom Control 61(7):1912–1917. https://doi.org/10.1109/TAC.2015.2482003

Kazantzis N, Kravaris C (1998) Nonlinear observer design using Lyapunov's auxiliary theorem. Syst Contr Lett 34:241–247

Keller H (1987) Non-linear observer design by transformation into a generalized observer canonical form. Int J Control 46:1915–1930

Kharitonov VL, Zhabko AP (2003) Lyapunov-Krasovskii approach to the robust stability analysis of time-delay systems. Automatica 39:15–20

Kim J, Chang PH, Park HS (2013) Two-channel transparency-optimized control architectures in bilateral teleoperation with time delay. IEEE Trans Control Syst Technol 62:2943–2953

Kotta Ü, Zinober ASI, Liu P (2001) Transfer equivalence and realization of nonlinear higher order input-output difference equations. Automatica 37:1771–1778

Krener AJ, Isidori A (1983) Linearization by output injection and nonlinear observers. Syst Contr Lett 3:47–52

Krener AJ, Respondek W (1985) Nonlinear observers with linearizable error dynamics. SIAM J Contr Opt 23:197–216

Krstic M (2009) Delay compensation for nonlinear, adaptive, and PDE. Systems Series: Systems & Control: Foundations & Applications. Birkhaüser, Boston

Krstic M, Bekiaris-Liberis N (2012) Control of nonlinear delay systems: a tutorial. 51st IEEE conference on decision and control (CDC). HI, Maui, pp 5200–5214

Lee EB, Olbrot A (1981) Observability and related structural results for linear hereditary systems. Int J Control 34:1061–1078

Li SJ, Moog CH, Califano C (2011) Characterization of accessibility for a class of nonlinear time-delay systems. In: CDC 2011. Orlando, pp 1068–1073

Li SJ, Califano C, Moog CH (2016) Characterization of the chained form with delays, IFAC NOL-COS. Monterey, CA, USA, 2011. IFAC-PapersOnLine 49-18:808–813

Mattioni M, Monaco S, Normand-Cyrot D (2018) Nonlinear discrete-time systems with delayed control: a reduction. Syst Control Lett 114:31–37

Mattioni M, Monaco S, Normand-Cyrot D (2021) IDA-PBC for LTI dynamics under input delays: a reduction approach. IEEE Control Syst Lett 5:1465–1470

Márquez-Martínez LA (1999) Note sur l'accessibilité des systèmes non linéaires à retards. Comptes Rendus de l'Académie des Sciences-Series I - Mathematics 329(6):545–550

Márquez-Martínez LA (2000). Analyse et commande de systèmes non linéaires à retards. PhD thesis, Université de Nantes / Ecole Centrale de Nantes, Nantes, France

Márquez-Martínez LA, Moog CH (2004) Input-output feedback linearization of time-delay systems. IEEE Trans Autom Control 49:781–786

Márquez-Martínez LA, Moog CH (2007) New insights on the analysis of nonlinear time-delay systems: application to the triangular equivalence. Syst Contr Lett 56:133–140

Márquez-Martínez LA, Moog CH, Velasco-Villa M (2002) Observability and observers for nonlinear systems with time delay. Kybernetika 38:445–456

Mazenc F, Bliman PA (2006) Backstepping design for time-delay nonlinear systems. IEEE Trans Autom Control 51:149–154

Mazenc F, Malisoff M, Krstic M (2021) Stability analysis using generalized sup-delay inequalities. IEEE Control Syst Lett 5:1411–1416

Mazenc F, Malisoff M, Lin Z (2008) Further results on input-to-state stability for nonlinear systems with delayed feedbacks. Automatica 44:2415–2421

Mazenc F, Malisoff M, Bhogaraju INS (2008) Sequential predictors for delay compensation for discrete time systems with time-varying delays. Automatica, 122, art n 109188. https://doi.org/10.1016/j.automatica.2020.109188

Michiels W, Niculescu S-I (2007) Stability and stabilization of time-delay systems. an eigenvalue-based approach. Advances in design and control, 12. Philadelphia, SIAM

Monaco S, Normand-Cyrot D (1984) On the realization of nonlinear discrete-time systems. Syst Control Lett 5:145–152

Monaco S, Normand-Cyrot D (2008) Controller and observer normal forms in discrete time. In: Isidori A, Astolfi A, Marconi L (eds) (in honor) Analysis and design of nonlinear control systems. Springer, pp 377–395

Monaco S, Normand-Cyrot D (2009) Linearization by output injection under approximate sampling. EJC 15:205–217

Monaco S, Normand-Cyrot D, Mattioni M (2017) Sampled-data stabilization of nonlinear dynamics with input delays through immersion and invariance. IEEE Trans Autom Control 62(5):2561–2567

Moraal PE, Grizzle JW (1995) Observer design for nonlinear systems with discrete-time measurement. IEEE Trans Autom Control 40:395–404

Moog CH, Castro-Linares R, Velasco-Villa M, Márquez-Martínez LA (2000) The disturbance decoupling for time-delay nonlinear systems. IEEE Trans Autom Control 45(2):305–309

Murray R, Sastry S (1993) Nonholonomic motion planning: steering using sinusoids. IEEE Trans Autom Control 38:700–716

Nijmeijer H, van der Schaft A (1990) Nonlinear dynamical control systems. Springer, New York

Niculescu SI (2001) Delay effects on stability: a robust control approach, vol 269. Lecture notes in control and information sciences, Springer, Heidelberg

Oguchi T (2007) A finite spectrum assignment for retarded non-linear systems and its solvability condition. Int J Control 80(6):898–907

Oguchi T, Watanabe A, Nakamizo T (2002) Input-output linearization of retarded non-linear systems by using an extension of Lie derivative. Int J Control 75:582–590

Olbrot AW (1972) On controllability of linear systems with time delay in control. IEEE Trans Autom Control 17:664–666

Olgac N, Sipahi R (2002) An exact method for the stability analysis of time-delayed linear time-invariant (LTI) systems. IEEE Trans Autom Control 47:793–797

Pepe P, Jiang ZP (2006) A Lyapunov-Krasovskii methodology for ISS and iISS of time-delay systems. Syst Control Lett 55:1006–1014

Plestan F, Glumineau A (1997) Linearization by generalized input-output injection. Syst Contr Lett 31:115–128

Richard JP (2003) Time-delay systems: an overview of some recent advances and open problems. Automatica 39(10):1667–1694

Sallet G (2008) Lobry Claude : un mathématicien militant. In: Proceedings of 2007 international conference in honor of Claude Lobry, ARIMA, 9, pp 5–13. http://arima.inria.fr/009/pdf/arima00902.pdf

Sename O, Lafay JF, Rabah R (1995) Controllability indices of linear systems with delays. Kybernetika 6:559–580

Shi P, Boukas EK, Agarwal RK (1999) Control of Markovian jump discrete-time systems with norm bounded uncertainty and unknown delay. IEEE Trans Autom Control 44:2139–2144

Sipahi R, Niculescu SI, Abdallah CT, Michiels W, Gu K (2011) Stability and stabilization of systems with time delay. IEEE Control Syst Mag 31(1):38–65

Sluis W, Shadwick W, Grossman R (1994) Nonlinear normal forms for driftless control systems. In: Proceedings of 1994 IEEE CDC, Lake Buena Vista, FL, 320–325

Sørdalen OJ (1993) Conversion of the kinematics of a car with n trailers into a chained form. In: Proceedings of 1993 international conference robotics and automation, Atlanta, CA, pp 382–387

Souleiman I, Glumineau A, Schreirer G (2003) Direct transformation of nonlinear systems into state affine MISO form and nonlinear observers design. IEEE Trans Autom Control 48:2191–2196

Spivak M (1999) A comprehensive introduction to differential geometry, 3rd edn. Publish or Perish, Houston

Timmer J, Müller T, Swameye I, Sandra O, Klingmüller U (2004) Modeling the nonlinear dynamics of cellular signal transduction. Int J Bifurc Chaos 14:2069–2079

Ushio T (1996) Limitation of delayed feedback control in nonlinear discrete-time systems. IEEE Trans Circuits Syst I(43):815–816

Van Assche V, Ahmed-Ali T, Hann CAB, Lamnabhi-Lagarrigue F (2011) High gain observer design for nonlinear systems with time varying delayed measurements. In: Proceedings 18th IFAC world congress, Milano, vol 44, pp 692–696

Velasco M, Alavarez JA, Castro R (1997) Disturbance decoupling for time delay systems. Asian J Control 7:847–864

Xia X-H, Gao W-B (1989) Nonlinear observer design by observer error linearization. SIAM J Cont Opt 27:199–216

Xia X, Márquez-Martínez LA, Zagalak P, Moog CH (2002) Analysis of nonlinear time-delay systems using modules over non-commutative rings. Automatica 38:1549–1555

Zhang H, Wang C, Wang G (2014) Finite-time stabilization for nonholonomic chained form systems with communication delay. J Robot Netw Artif Life 1:39–44

Zheng G, Barbot JP, Boutat D, Floquet T, Richard JP (2011) On observation of time-delay systems with unknown inputs. IEEE Trans Autom Control 56(8):1973–1978

Zheng G, Barbot J-P, Boutat D (2013) Identification of the delay parameter for nonlinear time-delay systems with unknown inputs. Automatica 49(6):1755–1760

Printed in the United States
by Baker & Taylor Publisher Services